Peter Bigler

NMR Spectroscopy: Processing Strategies

VCH
A Wiley company

Spectroscopic Techniques:
An Interactive Course

Pretsch/Clerc
Spectra Interpretation of Organic Compounds

Bigler
NMR Spectroscopy: Processing Strategies

In Preparation:

Weber/Thiele/Hägele
NMR Spectroscopy: **Modern Spectral Analysis**

Jenny
NMR Spectroscopy: **Data Acquisition**

Fröhlich
NMR Spectroscopy: **Intelligent Data Management**

Peter Bigler

NMR Spectroscopy: Processing Strategies

VCH

A Wiley company

Dr. Peter Bigler
Department of Chemistry and Biochemistry
University of Berne
Freiestraße 3
CH-3012 Bern
Switzerland

A CD-ROM containing a teaching version of the program WIN-NMR (© Bruker Analytik GmbH) is included with this book. Readers can obtain further information on this software by contacting: Bruker Analytik GmbH, Silberstreifen, D-76287 Rheinstetten, Germany.

Editorial Directors: Dr. Christina Dyllick, Cornelia Crauß
Production Manager: Peter J. Biel

Library of Congress Card No. applied for
A catalogue record for this book is available from the British Library

Die Deutsche Bibliothek – CIP-Einheitsaufnahme
NMR Spectroscopy: Processing Strategies / Peter Bigler. - Weinheim ; New York ; Basel ; Cambridge ; Tokyo ; VCH
(Spectroscopis techniques)
ISBN 3-527-28812-0
NE: Bigler, Peter

Buch. - 1997

CD-ROM. - 1997

ISBN 3-527-28812-0

Printed on acid-free and low chlorine paper

Composition: Kühn & Weyh, D-79111 Freiburg
Printing: Betzdruck GmbH, D-64291 Darmstadt
Bookbinding: Schäffer GmbH & Co. KG, D-67269 Grünstadt

Printed in the Federal Republic of Germany

Preface

High resolution NMR spectroscopy is currently the most popular technique in unravelling molecular structures. The main reason for this popularity are the various interactions between nuclei which may be detected and determined quantitatively by corresponding NMR experiments. Whether the aim is to elucidate the structure of an unknown pure compound, to measure proton-proton distances in a protein or to detect and quantify the signals of metabolites from a biological extract, it is those properties relating one nucleus with another, which makes NMR such an indispensable tool not only in chemistry but also in biology, medicine and related sciences. As a consequence, numerous pulse experiments have been designed to exploit these nuclear interactions and as a result the structural information now available with high resolution NMR spectroscopy is probably greater and more readily obtainable than with any other single technique.

Over the last few years there has been a tremendous technical improvement in NMR spectrometer design. The increasing number of modern research and low-cost FT NMR spectrometers and the powerful NMR software available today have lead to new areas of application and new perspectives of how to use and exploit NMR spectroscopy. New concepts have to be introduced and proved which should not only include the maximisation of sample through put, but should also encourage NMR users to undertake part of the tasks usually done exclusively by the NMR specialist. Reassigning the various jobs among users and specialists, taking into consideration the users and the specialists theoretical background and NMR expertise, should increase overall efficiency and bring the beauties of modern NMR closer to the interested user. This reassigning of responsibilities can take two forms,

- Routine NMR spectrometers can be operated either in automation mode or "hands-on" mode by specially trained users, allowing the specialist NMR operators to concentrate on more demanding spectroscopic problems.
- The enormous amount of NMR raw data produced by a modern spectrometer can be processed on remote computers. The power and capacity of even low-cost personal computers, the versality of corresponding NMR software and the availability of local networks for rapid data transfer allow the non-specialised user to efficiently process and analyse their own NMR data on a remote computer station. This will increase the sample through put and give the NMR specialists more time to use the spectrometer computer for testing and optimising new sophisticated experiments or to do time-consuming and more demanding processing.

These ideas and perspectives were the origin for the series entitled ***Spectroscopic Techniques: An Interactive Course***. The section relating to ***NMR Spectroscopy***, consists of four volumes

- Volume 1 – *Processing Strategies*
- Volume 2 – *Data Acquisition*
- Volume 3 – *Modern Spectral Analysis*
- Volume 4 – *Intelligent Data Management*

and deals with all the aspects of a standard NMR investigation, starting with the definition of the structural problem and ending – hopefully – with the unravelled structure. This sequence of events is depicted on the next page. The central step is the transformation of the acquired raw data into a NMR spectrum which may then be used in two different ways. The NMR spectrum can be analysed and the NMR parameters such as chemical shifts, coupling constants, peak areas (for proton spectra) and relaxation times can be extracted. Using NMR parameter data bases and dedicated software tools these parameters may then be translated into structural information. The second way follows the strategy of building up and making use of NMR data bases. NMR spectra serve as the input for such data bases, which are used to directly compare the measured spectrum of an unknown compound either with the spectra of known compounds or with the spectra predicted for the expected chemical structure. Which of the two approaches is followed depends on the actual structural problem. Each of them has ist own advantages, limitations and filed of application. However, it is the combined application of both techniques that makes them such a powerful tool for structure elucidation.

The contents of volumes 1–4 may be summarized as follows:

Volume 1: ***Processing Strategies***

Processing NMR data transforms the acquired time domain signal(s) – depending on the experiment – into 1D or 2D spectra. This is certainly the most central and important step in the whole NMR analysis and is probably the part, which is of interest to the vast majority of NMR users. Not everyone has direct access to an NMR spectrometer, but most have access to some remote computer and would prefer to process their own data according to their special needs with respect to their spectroscopic or structural problem and their ideas concerning the graphical layout *i.e.* for presentation of reports, papers or thesis. It is essential for the reliability of the extracted information and subsequent conclusions with respect to molecular structure, that a few general rules are followed when processing NMR data. It is of great advantage that the user is informed about the many possibilities for data manipulation so they can make the best use of their NMR data. This is especially true in more demanding situations when dealing with subtle, but nevertheless important spectral effects. Modern NMR data processing is not simply a Fourier transformation in one or two dimensions, it consists of a series of additional steps in both the time and the frequency domain designed to improve and enhance the quality of the spectra.

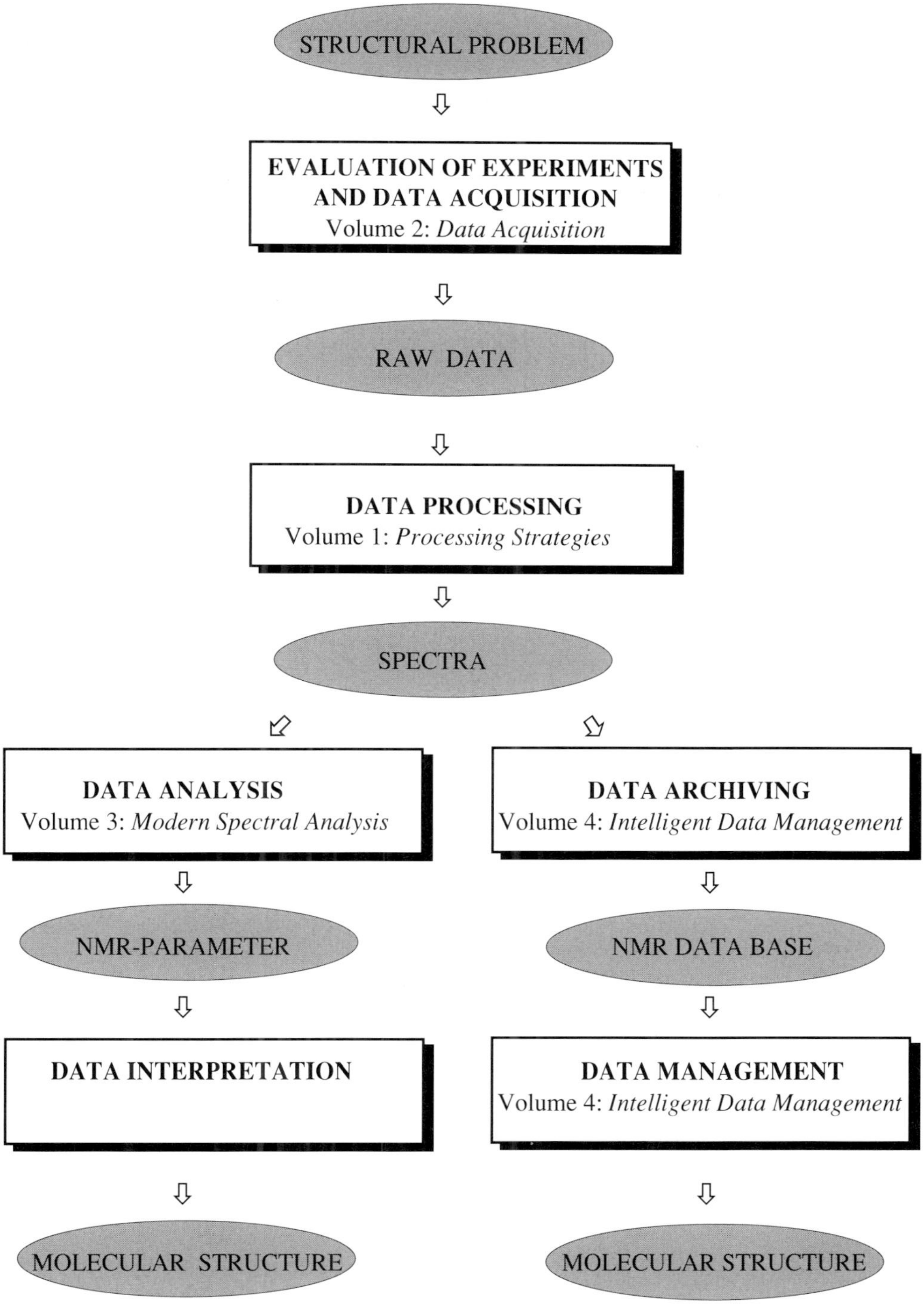
STRUCTURAL PROBLEM
EVALUATION OF EXPERIMENTS AND DATA ACQUISITION
Volume 2: *Data Acquisition*
RAW DATA
DATA PROCESSING
Volume 1: *Processing Strategies*
SPECTRA
DATA ANALYSIS
Volume 3: *Modern Spectral Analysis*
DATA ARCHIVING
Volume 4: *Intelligent Data Management*
NMR-PARAMETER
NMR DATA BASE
DATA INTERPRETATION
DATA MANAGEMENT
Volume 4: *Intelligent Data Management*
MOLECULAR STRUCTURE
MOLECULAR STRUCTURE

Processing Strategies gives the theoretical background for all these individual processing steps and demonstrates the effects of the various manipulations on suitable examples. The powerful Bruker 1D WIN-NMR, 2D WIN-NMR and GETFILE software tools, together with a set of experimental data for two carbohydrate compounds allow you to carry out the processing steps on your own remote computer, which behaves in some sense as a personal "NMR processing station". You will learn how the quality of NMR spectra may be improved, experience the advantages and limitations of the various processing possibilities and most important, as you work through the text, become an expert in this field. The unknown structure of one of the carbohydrate compounds should stimulate you to exercise and apply what you have learnt. The elucidation of this unknown structure should demonstrate, how powerful the combined application of several modern NMR experiments can be and what an enormous and unexpected amount of structural information can thereby be obtained and extracted by appropriate data processing. It is this unknown structure which should remind you throughout this whole educational series that NMR data processing is neither just "playing around" on a computer nor some kind of scientific "l'art pour l' art". The main goal for measuring and processing NMR data and for extracting the structural information contained in it, is to get an insight into how molecules behave.

Furthermore, working through ***Processing Strategies*** should encourage you to study other topics covered by related volumes in this series. This is particularly important if you intend to operate a NMR spectrometer yourself, or want to become familiar with additional powerful software tools to make the best of your NMR data.

Volume 2: ***Data Acquisition***

Any NMR analysis of a structural problem usually starts with the selection of the most appropriate pulse experiment(s). Understanding the basic principles of the most common experiments and being aware of the dependence of spectral quality on the various experimental parameters are the main prerequisites for the successful application of any NMR experiment. Spectral quality on the other hand strongly determines the reliability of the structural information extracted in subsequent steps of the NMR analysis. Even if you do not intend to operate a spectrometer yourself, it would be beneficial to acquire some familiarity with the interdependence of various experimental parameters e.g. acquisition time and resolution, repetition rate, relaxation times and signal intensities. Many mistakes made with the application of modern NMR spectroscopy arise because of a lack of understanding of these basic principles. ***Data Acquisition*** covers these various aspects and exploits them in an interactive way using the Bruker software package NMRSIM. Together with 1D WIN-NMR and 2D WIN-NMR, NMRSIM allows you to simulate routine NMR experiments and to study the interdependence of a number of NMR parameters and to get an insight into how modern multiple pulse NMR experiments work.

Volume 3: ***Modern Spectral Analysis***

Following the strategy of spectral analysis, the evaluation of a whole unknown structure, of the local stereochemistry in a molecular fragment or of a molecules dynamic properties, depends on NMR parameters. Structural informations are obtained in subsequent steps from chemical shifts, homo- and heteronuclear spin-spin connectivities and corresponding coupling constants and from relaxation data such as NOEs, ROEs, T_1s or T_2s and assumes that the user is aware of the typical ranges of these NMR parameters and of the numerous correlations between NMR and structural parameters, i.e. between coupling constants, NOE enhancements or linewidths and dihedral angles, internuclear distances and exchange rates respectively. However, the extraction of these NMR parameters from the corresponding spectra is not always straightforward,

- The spectrum may exhibit extensive signal overlap, a problem common with biomolecules.
- The spectrum may contain strongly coupled spin systems.
- The molecule under investigation may be undergoing dynamic or chemical exchange.

Modern Spectral Analysis discusses the strategies needed to efficiently and competently extract NMR parameters from the corresponding spectra. You will be shown how to use the spectrum simulation package WIN-DAISY to extract chemical shifts, coupling constants and individual linewidths from even highly complex NMR spectra. In addition, the determination of T_1s, T_2s or NOEs using the special analysis tools of 1D WIN-NMR will be explained. Sets of spectral data for a series of representative compounds, including the two carbohydrates mentioned in volume 1 are used as instructive examples and for problem solving. NMR analysis often stops with the plotting of the spectrum thereby renouncing a wealth of structural data. This part of the series should encourage you to go further and fully exploit the valuable information "hidden" in the carefully determined NMR parameters of your molecule.

Volume 4: ***Intelligent Data Management***

The evaluation and interpretation of NMR parameters to establish molecular structures is usually a tedious task. An alternative way to elucidate a molecular structure is to directly compare its measured NMR spectrum – serving here as a fingerprint of the investigated molecule – with the corresponding spectra of known compounds. An expert system combining a comprehensive data base of NMR spectra with associated structures, NMR spectra prediction and structure generators not only facilitates this part of the NMR analysis but makes structure elucidation more reliable and efficient.

In ***Intelligent Data Management***, an introduction to the computer-assisted interpretation of molecular spectra of organic compounds using the Bruker WIN-SPECEDIT software package is given. This expert system together with the Bruker STRUKED software tool is designed to follow up the traditional processing of NMR spectra using 1D- and 2D WIN-NMR in terms of structure-oriented spectral interpretation and signal assignments. WIN-SPECEDIT offers not only various tools for automatic interpretation of spectra and for structure elucidation, including the prediction of spectra, but also a number of functions for so-called „authentic" archiving of spectra in a database, which links molecular structures, shift information and assignments with original spectroscopic data. You will learn to exploit several interactive functions such as the simple assignment of individual resonances to specific atoms in a structure and about a number of automated functions such as the recognition of signal groups (multiplets) in ^{1}H NMR spectra. In addition, you will also learn how to calculate and predict chemical shifts and how to generate a local database dedicated to your own purposes. Several examples and exercises, including the two carbohydrate compounds, serve to apply all these tools and to give you the necessary practice for your daily spectroscopic work.

It is the primary aim of the series to teach the user how NMR spectra may be obtained from the data acquired on a spectrometer and how these spectra may be used to establish molecular structure following one of the two strategies outlined before. The series of volumes therefore emphasises the methodical aspect of NMR spectroscopy, rather than the more usual analytical aspects *i.e.* the description of the various NMR parameters and of how they depend on structural features, presented in numerous text books.

This series of books is to give the newcomer to physical NMR spectroscopy the necessary information, the theoretical background and the practice to acquire NMR spectra, to process the measured raw data from modern routine homo- and heteronuclear 1D and 2D NMR experiments, to evaluate NMR parameters, to generate and exploit dedicated data bases and finally to establish molecular structures.
Each of the four volume consists of three parts:

- A written part covers the theoretical background and explains why things are done in particular manner. Practical hints, examples, exercises and problems are also included.
- Software tools dedicated to the items discussed in the corresponding volume are supplied on CD-ROM.
- The most popular 1D and 2D pulse sequences together with the corresponding NMR raw data and spectra are supplied on CD-ROM. They are used to simulate NMR experiments, to exercise data processing and spectral analysis and serve as a data base for spectral interpretation.

It is this combination of written text, the software tools and data supplied, that make it different from other books on NMR spectroscopy and which should draw your attention to the many possibilities and the enormous potential of modern NMR. Sitting in front of your PC , which becomes your personal “PC-NMR spectrometer”, you experience in a very direct and practical way, how modern NMR works. According to the approved rule “Learning by Doing” you perform NMR experiments without wasting valuable spectrometer time, handle experimental data in different ways, plot 1D and 2D spectra, analyse spectra and extract NMR parameters and learn to build up and use NMR data bases.

 ⇔ 

TEXTBOOK **PC**

TEXTBOOK	PC
THEORY	SOFTWARE TOOLS
PRACTICAL HINTS	PULSE SEQUENCES
EXERCISES	NMR DATA
PROBLEMS	

It is recommended that you use all these educational tools in a complementary and interactive way switching from textbook to the software tools and the sets of data stored on the PC and back again and that you proceed at your own rate. It is assumed that you verify the numerous examples and solve the exercises in order to improve your skill in using the various software tools and to consolidate the theoretical background. In this way, the strongly interconnected components of this series of books are best utilised and will guarantee the most efficient means to become an expert in this field.

Furthermore it is recommended that NMR newcomers start with the central volume ***Processing Strategies*** and complete their education in modern NMR spectroscopy according to their special needs by working through the appropriate volumes, ***Data Acquisition***, ***Modern Data Analysis*** and ***Intelligent Data Management***.

This interactive course in practical NMR spectroscopy may be used in dedicated courses in modern NMR spectroscopy at universities, technical schools or in industry, or may be used in an autodidactic way for those interested in this field.

Acknowledgements

I am deeply indebted to Dr. B. F. Taylor, University of Sheffield, for checking and proofreading the entire manuscript, for many valuable comments and his encouragement in preparing this volume.

I am very grateful to Dr H. Thiele, Dr. A. Germanus and Dipl. Ing. J. Skarbek (BRUKER-Franzen Analytik) who developed the WIN-NMR software modules for their helpful advices and the excellent collaboration.

I would also like to express my gratitude to BRUKER/SPECTROSPIN for their interest in this project, helpful advice and support, and to Wiley-VCH for their assistance and patience when waiting for the final manuscript.

Finally I thank my family and my research group who had to put up with far less attention than they deserved, for much longer than they, or I, expected.

Table of Contents

1 Introduction

1.1 Scope and Audience

Information about the structure of molecules may be obtained by the application of appropriate NMR experiments and may be extracted from the corresponding spectra after appropriate data processing (Fig. 1.1). In a primary step experimental raw data, the so-called free induction decay (FID), is sampled and digitally stored. NMR data processing transforms this complex and unreadable raw data into spectra, which may be analysed and interpreted by the spectroscopist. In the course of this stepwise transformation NMR data may be modified and manipulated in many different ways. Among the various tasks associated with NMR data processing such as calibrating and referencing spectra, the production of high quality spectra is always of paramount importance. This influences directly the quality of NMR data archives if the spectra are used as inputs and it finally determines the precision and reliability of the desired NMR parameters (chemical shifts, coupling constants, relaxation or exchange parameters) which are extracted from the spectra in the subsequent steps of a NMR analysis. NMR processing is therefore certainly not a "quantité négligeable" but is an important factor which influences the quality of an NMR spectroscopic structural analysis in general.

Processing Strategies was designed to introduce newcomers in the fascinating field of NMR into the important and central step of NMR data processing. Numerous books have been published dealing with the more "analytical" aspect of NMR spectroscopy such as the theory of chemical shifts and coupling constants and their dependence on structural parameters *etc.* but very little has been presented dealing with this more "methodical" aspect of NMR spectroscopy. It is the purpose of this book to help the non-specialist go beyond the normal analytical aspect and to enter this more methodical part of NMR spectroscopy. It should enable and encourage you to process your data according to your own special needs and ideas rather than have to rely on automatic processing or specialist help.

Processing Strategies gives you the necessary information and practice to become competent in the processing of NMR data. It gives an overview of valuable software tools, their advantages, disadvantages and limitations to enable the best results to be obtained from the data available. *Processing Strategies* is therefore invaluable to any newcomer who wishes to enter into the methodical realm of NMR spectroscopy.

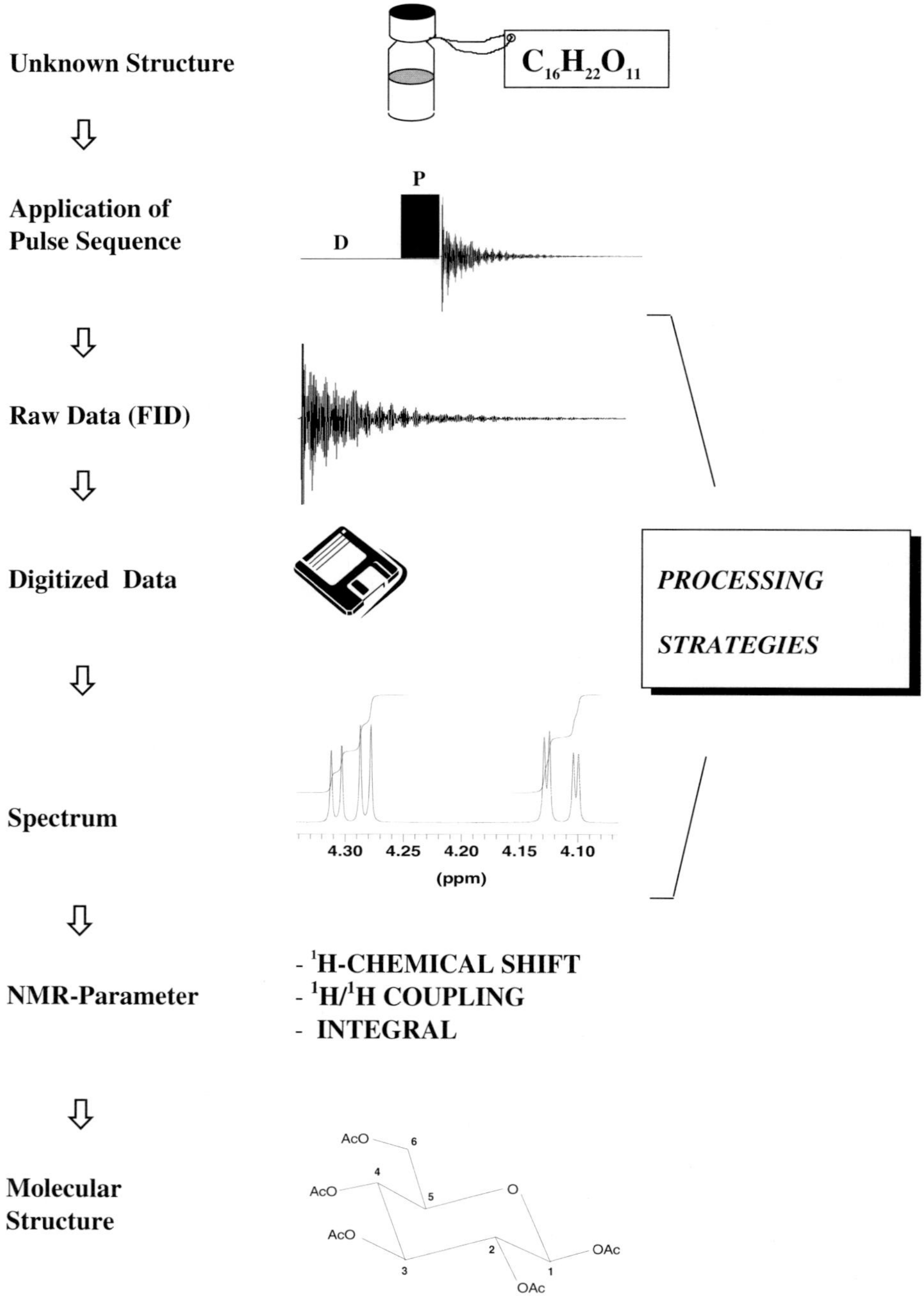

Fig. 1.1: Basic steps of an NMR investigation based on spectral analysis

1.2 Organisation

Processing Strategies is composed of three sections (Fig. 1.2):

- A written text giving an introduction into the basic theory and presenting practical hints, examples and exercises (*Check its*) to directly apply what you have learnt in practice.
- Software tools including manuals (HELP routines) describing in detail the use of 1D WIN-NMR, 2D WIN-NMR and GETFILE supplied on CD-ROM.
- A data base consisting of the NMR data for a series of the most popular and modern NMR experiments applied to two carbohydrate compounds supplied on CD-ROM.

PROCESSING STRATEGIES		Basic Theory, Rules, Hints, Recommendations Examples, Exercises
WIN-NMR SOFTWARE		
	1D WIN-NMR	Processing of 1D NMR Data HELP Routine
	2D WIN-NMR	Processing of 2D NMR Data HELP Routine
	GETFILE	Import/Export of NMR data via local networks and conversion of the original data format to the WIN-NMR format
NMR DATA BASE		
	β-D-Glucose	1D and 2D NMR Data (FID, Spectra)
	Oligosaccharide	1D and 2D NMR Data (FID)

Fig. 1.2: Components of *Processing Strategies* (volume 1)

These three sections are combined in such a way that you may use all these educational tools interactively with your PC. In this way as you learn about the basics in theory, you directly apply what you have learnt in practice following the instructions given in the numerous *Check its*. You experience step by step the scope of the powerful software modules, their advantages and limitations and you acquire the necessary skills to become an expert in NMR data processing. It is essential at this stage to exploit and make use of the powerful HELP routines if necessary. In contrast to the introductory text, discussing the effects and the purpose of the various processing options, these HELP routines available with the WIN-NMR software modules give detailed informations for how to apply them in practice.

The comprehensive 1D and 2D NMR data obtained from various modern NMR experiments within the data base serve two main purposes:

1. The FIDs measured for the first carbohydrate compound (peracetylated β-D-glucose) are used as examples demonstrating the various processing options and their effects on the resulting spectra. This allows you to perform the same processing steps yourself and to verify and compare your results with the corresponding spectra stored in the same data base. These spectra and the corresponding NMR parameters may furthermore be used as a spectral reference when dealing with the NMR data of the second carbohydrate of unknown structure.
2. The NMR data of the second carbohydrate (peracteylated oligosaccharide) consists of the FIDs only. As an exercise you should transform this data following the rules, recommendations and hints presented in this book. Applying your own ideas you should plot the corresponding spectra and try to unravel the unknown structure of this compound.

In the course of these exercises you should be stimulated to work through the additional volumes in this series either to learn more about the application of modern NMR experiments (*Data Acquisition* - volume 2), about extracting NMR parameters (*Modern Spectral Analysis* - volume 3) or about using NMR spectra data bases to unravel unknown molecular structures (*Intelligent Data Management* - volume 4).

Both the NMR spectra of two carbohydrates and the structural informations extracted from these spectra will hopefully show you, how beautiful and valuable today's NMR really is. This should encourage you to apply powerful NMR experiments and to exploit all facets of modern NMR spectroscopy for solving your own structural problems.

1.3 Personal Qualifications

It is assumed that you know the basic theory of Pulse Fourier Transform NMR, and that you have a basic understanding of simple pulse experiments (see Recommended Reading at the end of this chapter). You should have completed at least one of the corresponding introductory courses usually given at universities or technical schools, where the various NMR parameters are introduced and their structure dependence is demonstrated. It is an advantage if you have some experience in analysing and interpreting simple NMR spectra. Furthermore you should be familiar with the use of a PC and the WINDOWS operating system. Although a few WINDOWS tools which may be especially helpful when applying the WIN-NMR software are outlined, you should be able to install software and to copy files either in the WINDOWS 3.1, WINDOWS 3.11 or the WINDOWS 95 environment. Consult the corresponding manuals of the various versions of the WINDOWS operating system if necessary.

1.4 Content

The text book is divided into six chapers:

1. **Introduction**
2. **Your Personal "NMR Processing Station"**
3. **Modern Homo- and Heteronuclear 1D- and 2D NMR Experiments**
4. **How to Display and Plot 1D- and 2D NMR Spectra**
5. **How to Process 1D- and 2D NMR Data**
6. **NMR Data of an Unknown Oligosaccharide**

Chapter 2 deals with your personal "NMR Processing Station", its technical requirements, the software and NMR data base supplied on CD-ROM and how to install it. It discusses importing data from remote computers, either the spectrometer computer itself or some file server and gives an example of how your PC could be connected to a local network so you may have daily access to your NMR data. It also gives useful information concerning the different Bruker data formats and briefly mentions how NMR data of other spectrometer manufacturers can be converted into a form where it may be processed with the WIN-NMR software modules using the GETFILE software.

Chapter 3 gives an overview of the most important of to-days NMR experiments applied to peracetylated β-D-glucose. The chapter starts with a general description of 1D and 2D NMR experiments. A number of pulse experiments are briefly described with their advantages and limitations; the kind of structural information that may be obtained is discussed. This information is a prerequisite to understanding the reasons why raw data from different experiments are processed differently and why the corresponding final spectra look different. For an extended and more detailed description, including the mechanics and the setting up of these experiments you are referred to *Data Acquisition* - volume 2.

Chapter 4 serves two purposes. First it allows you to acquire your first experience with the software tools 1D WIN-NMR and 2D WIN-NMR and how they are used to process, display and plot 1D and 2D spectra. These tasks are in many cases the first and most regularly performed steps when starting to process NMR on a "Do it Yourself" basis. It is important to obtain the best possible spectrum to facilitate the subsequent extraction of NMR parameters as discussed in *Modern Data Analysis* - volume 3. 1D WIN-NMR and 2D WIN-NMR offer a variety of features for final processing, displaying and plotting spectra. These include the extremes of a facility to quickly inspect spectra when monitoring a reaction or on a more sophisticated level to design your final layout to be included, in a publication or thesis. It is the aim to give you an overview of what is possible without great detail, since the relevant information is contained in the corresponding HELP routines supplied on CD-ROM.

Secondly this chapter serves to improve your skill in performing these tasks for a whole series of spectra obtained with the experiments discussed in chapter 3 for peracetylated β–D–glucose. It is therefore highly recommended that you display and plot the corresponding spectra as you read the text. This will improve your data processing abilities and will illustrate what modern NMR experiments can do and what kind of structural information may be obtained from the different spectra.

The spectroscopic parameters extracted from the glucose spectra will also serve as a valuable reference to elucidate the unknown structure of the oligosaccharide, which is the subject of chapter 6.

Last but not least the familiarity with displaying and plotting 1D- and 2D NMR spectra is a prerequisite to inspect, understand and evaluate the results obtained with the many processing options discussed in chapter 5.

Chapter 5 is the most exciting and important part of this book and deals with all aspects of modern NMR data processing. In the first section the general scheme for processing 1D and 2D NMR data and the main steps in transforming the FID into a spectrum are outlined. Each section in this chapter is organized in a similar way. After a short introduction, explaining the theoretical background and the reasons for applying a particular processing option, practical advice is given on how to use the option together with examples demonstrating its effect on the final spectrum. These examples and corresponding *Check its* allow you to verify what has been discussed and to acquire the necessary practical skills in applying a particular processing option. For these purposes, the raw data and the spectra of peracetylated β–D–glucose in the CD-ROM data base are used.

Later sections of this chapter deal with more advanced and specialised processing options such as zero filling, linear prediction, deconvolution and the manipulation of 2D data sets. The chapter concludes with a set of tables containing recommendations for the type of processing function and the corresponding parameters to be used in a number of 1D and 2D experiments.

Chapter 6 deals exclusively with the determination of the structure of an "unknown" oligosaccharide and is designed to test both your skill and understanding of data processing. The chapter starts with a discussion of the different types of strategies that may be used in structural determination. The spectroscopic data of the known glucose used in the *Check its* in chapters 4 and 5 and a section on the NMR parameters of carbohydrates provide a source of reference data. The *Check its* allow the processing of the data of this peracetylated oligosaccharide in an analogous manner to the known peracteylated glucose. The last section shows you what kind of information can be extracted from the processed spectra and how it can be combined to elucidate the structure of this unknown carbohydrate before finally revealing the structure.

1.5 Recommended Reading

Akitt, J. W. *NMR and Chemistry: An Introduction to NMR Spectroscopy*, 3rd ed Chapman & Hall, 1992, published in UK.

Farrar, T. C., *Introduction to Pulse NMR Spectroscopy*, 2nd Farragut Press, 1989, published in USA.

Günther, H., *NMR Spectroscopy*, 2nd ed John Wiley Ltd, 1995, published in UK.

Hoult, D. I., *The Magnetic Resonance Myth of Radio Waves*, Concepts in Magnetic Resonance; An Educational Journal, 1989, *1* (No 1)

Sanders, J. K. M., Constable, E. C., Hunter, B. K., Pearce, C., *Modern NMR Spectroscopy: A Guide for Chemists: A Workbook of Chemical Problems*, 2nd ed Oxford UP, 1993, publsihed in UK.

Sanders, J. K. M., Hunter, B. K., *Modern NMR Spectroscopy: A Guide for Chemists*, 2nd ed Oxford UP, 1993, published in UK.

2 Your Personal "PC-NMR-Processing Station"

2.1 Introduction

This chapter contains information on the hardware and software requirements for installing and running the software tools 1D WIN-NMR and 2D WIN-NMR and for installing the NMR data base used with this and other volumes of *Spectroscopic Techniques: An Interactive Course.* For later applications, i.e. the processing of your personal NMR data using the standard WIN-NMR programs, this chapter contains additional information to allow you to configure your PC as a powerful system optimised to your requirements. This chapter describes the origin, the content and the directory structure of the experimental NMR data base that is supplied on CD-ROM. It also gives an overview of the various formats of Bruker NMR data files and discusses the conversion of these different data formats – including the data files of other spectrometer manufacturers – into the format used with WIN-NMR. Finally an example of how to connect your PC to a local network for data import from remote computers is presented.

2.2 Technical Requirements

In order to install and run 1D WIN-NMR, 2D WIN-NMR and GETFILE you will need an IBM compatible 386, 486 or Pentium class PC with a minimum of 4 MB RAM base memory. An arithmetic coprocessor is needed if a corresponding unit for floating point operations is not integrated in the CPU. All graphics cards and output devices for which MS-WINDOWS drivers are available are supported by the programs, but a graphic card and a monitor which allows at least standard VGA resolution (640 by 480) are recommended. A pointing device that is compatible with the Microsoft two button mouse is needed to run the two WIN-NMR programs and the GETFILE program.

The software and the comprehensive NMR data base are both stored on a CD-ROM. A corresponding drive is therefore required and a hard disk with at least 100 MB spare capacity is advisable.The use of floppy diskettes for importing 1D NMR data sets is possible, but this is not feasible for 2D data sets because of their size.

If you plan to use your PC for accessing NMR data stored on remote computers, you will need additional hardware and software (see below). Your PC should also be connected to a local network.

2.3 Software Tools

2.3.1 General

1D WIN-NMR and 2D WIN-NMR have been developed to process 1D NMR and 2D NMR data on a PC under the MS-WINDOWS environment. The 2D WIN-NMR program is also linked to the 1D WIN-NMR program for special 2D processing. The programs can process data generated on various Bruker NMR spectrometers (AC, AM, CPX, MSL, AMX, ARX, DPX, DRX, DMX,) as well as data produced by one of Bruker's MS-WINDOWS based simulation programs. Furthermore NMR data acquired with spectrometers of other manufacturers (Varian, GE, JEOL) may be converted and processed as well.

To run 1D WIN-NMR and 2D WIN-NMR the MS-DOS operating system version 3.3. or higher and MS-WINDOWS version 3.1 or higher, including the WIN-COMMANDER are needed. Using the powerful MS-WINDOWS NT operating system offers additional options and versality with the 1D and 2D WIN-NMR software. For any further details including the installation of these operating systems the reader is referred to the corresponding software manuals.

The volumes in the series of *Spectroscopic Techniques: An Interactive Course* are delivered with special versions of 1D WIN-NMR and 2D WIN-NMR. They are a supplement for this course to be installed on a stand-alone PC and to be used exclusively for processing the experimental data supplied in the NMR data base. They cannot be used to process the users personal NMR data. The full version of 1D WIN-NMR and 2D WIN-NMR software must be installed for this purpose and a special copy protection dongle (a WIBU key for the single user mode, or a Net-HASP key for the multi-user/network mode) must be used. Note also that for 2D WIN-NMR a standard 16-bit and a more powerful 32-bit version exist. Please refer to the description in the corresponding Bruker manuals [2.1, 2.2].

The GETFILE software is used in this volume to demonstrate its capabilities for converting various spectrometer specific data formats into the WINNMR format. It has been developed for data conversion and easy data transfer and must be installed if you wish to

1. Process NMR data measured either on older Bruker spectrometers (AC, AM), connected to ASPECT computers using the ADAKOS/DISNMR software, or measured on spectrometers of other manufacturers.
2. If your PC is connected to a local network and you want to have direct access to NMR data stored on remote computers.

Contact your local Bruker representative or Bruker-Franzen Analytik GmbH in Bremen (Germany), for the actual versions of the two standard WIN-NMR programs and the GETFILE program and for additional information concerning the different dongle types and for detailed descriptions of how to install and operate these software tools, if you want to use them for your future spectroscopic work.

To install 1D WIN-NMR, 2D WIN-NMR and GETFILE, stored in the directory NMRPROG on your CD-ROM, you must first have installed the MS-DOS operating system version 3.3 or higher and MS-WINDOWS version 3.1 or higher on your PC. Please refer to the documentation that came with these products for how to do this.
Note: The installation and the starting of the software tools (2.3.2 - 2.3.5), the copy of the NMR data (2.5.5) and the description of a few useful WINDOWS options (2.5.6) is demonstrated for a PC under a MS-WINDOWS 95 environment. However the corresponding operations and options are also available with MS-WINDOWS 3.1 or 3.11.

2.3.2 Installation of 1D WIN-NMR

In the following instructions describing the installation of the educational version of 1D WIN-NMR (1D WIN-NMR Demo) it is assumed that these software modules are stored in a subdirectory E:\NMRPROG\WIN1D on your CD-ROM and that you want to install it in a subdirectory on disk C: of your PC. Note that for this educational version a directory DEMO1D is created on your harddisk to avoid any problems with the full version of 1D WIN-NMR which you will eventually use for your later work and which will be stored in a directory WIN1D. Note also that with the full version installation on other disks (e.g. D:) is also possible.

Check it in WINDOWS:

Insert the CD-ROM into the drive and follow the standard procedure to install a program in the WINDOWS environment. Consult the MS-WINDOWS manual if necessary. Check there is 10 MBytes of space available on disk C:. Select the set-up program for 1D WIN-NMR Demo, E:\NMRPROG\WIN1D\SETUP. EXE, on the CD-ROM and initiate the set-up process. After a few seconds the 1D WIN-NMR Setup window will appear on the screen. Click the **Continue** button to proceed with the installation of 1D WIN-NMR which will open a series of dialog boxes one after the other. Confirm the (default) directories in the second 1D WIN-NMR Installation dialog box and select the options for installing 1D WIN-NMR, for installing the example files and for creating a program item. Note, however, that with other disk drives available on your PC, and depending on which drive will be selected when copying the NMR data directory NMRDATA from the CD to your PC (see 2.5.5), the Spectra directory may also be positioned on another drive, e.g. disk D: as shown in Fig. 2.1. Do not use the **RETURN** key to confirm or terminate the individual entries as this would immediately start the installation process. When all entries fit your needs click on the **Continue** button to leave this dialog box and to initialize the installation. The corresponding message appears on screen after successful installation of 1D WIN-NMR and the Program Manager WIN-NMR group is created, which contains the icon for the 1D WIN-NMR Demo program. You can run 1D WIN-NMR Demo from Program Manager by double clicking on this icon (see 2.3.5).

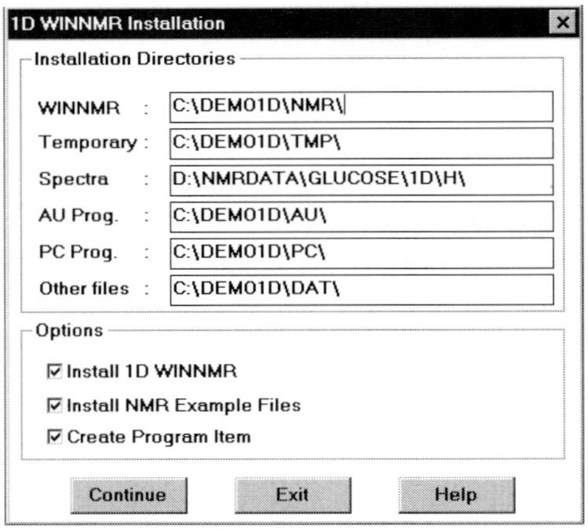

Fig. 2.1: 1D WIN-NMR installation box. Note that the optional disk D: has been chosen for the Spectra directory.

Note: During the installation of the educational versions of 1D WIN-NMR and 2D WIN-NMR three files, DEMO1D.INI, DEMO2D.INI and BRUKERDE.INI are generated in the WINDOWS directory on your PC. Various sections in the DEMO1D.INI file allow you to tailor 1D WIN-NMR to your personal hardware situation (printer, plotter) and to your personal needs and preferences and to set the corresponding default values, defining directories used by 1D WIN-NMR, units and colors in your display, the thickness of the lines used for plotting, the various directories initially accessed to display and copy lists, metafiles and others, to set upper limits for the number of peaks in your peak list or for the number of intervals used during integration and many others. The DEMO2D.INI file consists of a series of sections made during the setup and additional settings not edited during setup, defining directories used by 2D WIN-NMR, comprising information on the size of the standard fonts for the various display fields on the screen, containing information on the automatic extrema calculation, specifying colors and many others. The BRUKERDE.INI file mainly contains information required for communication between the various programs which form the Bruker WINDOWS family. The three files DEMO1D.INI, DEMO2D.INI and BRUKERDE.INI which correspond to WINNMR-1D.INI, D2NMR.INI and BRUKER.INI in the full version of 1D WIN-NMR and 2D WIN-NMR respectively, are described in detail in the 1D WIN-NMR [2.1] and 2D WIN-NMR [2.2] manuals.

2.3.3 Installation of 2D WIN-NMR

In the following instructions describing the installation of the educational version of 2D WIN-NMR (2D WIN-NMR Demo) it is again assumed that these software modules are stored in a subdirectory E:\NMRPROG\ WIN2D on your CD-ROM and that you want to install it in a subdirectory on disk C: of your PC. Note that for this educational version the standard 16-bit rather than the more powerful 32-bit version will be used and a directory DEMO2D will be created on your harddisk to avoid any problems with the full version of 2D WIN-NMR which you will eventually use for your later work and which will be stored in a directory WINNMR2D. Note also that with the full version installation on other disks (e.g. D:) is also possible.

Follow again the standard procedure to install a program in the WINDOWS environment. Consult the MS-WINDOWS manual if necessary. Check there is 10 MBytes space available on disk C:. Select the set-up program for 2D WIN-NMR, E:\NMRPROG\WIN2D\SETUP.EXE, on the CD-ROM and initiate the set-up process. After a few seconds the 2D WIN-NMR Setup window will appear on the screen. Click the **Continue** button to proceed with the installation of 2D WIN-NMR which will open a series of dialog boxes one after the other. Select the **Single User Installation** option and the **16 Bit** version. Confirm the (default) directories shown in the next 2D WIN-NMR Installation dialog boxes and the **Default cutting level** and select **no** for the **Automatic cutting level** option. Note, however, that with other disk drives available on your PC, and depending on which drive will be selected when copying the NMR data directory NMRDATA from the CD to your PC (see 2.5.5), the System and the 1D-Spectra directories may also be positioned on another drive, e.g. disk D: as shown in Fig. 2.2. Use the **Accept** button in the last dialog box to confirm the entries for the various directories at the end of the interactive installation. Do not use the **RETURN** key to confirm or terminate the individual entries as this would immediately start the installation process. When all entries fit your needs click on the **Continue** button to leave this dialog box and to initialize the installation. A corresponding message appears on screen after successful installation of 2D WIN-NMR and the Program Manager WIN-NMR group is created, which contains the icon for the 2D WIN-NMR Demo program. You can run 2D WIN-NMR Demo from Program Manager by double clicking on this icon (see 2.3.5).

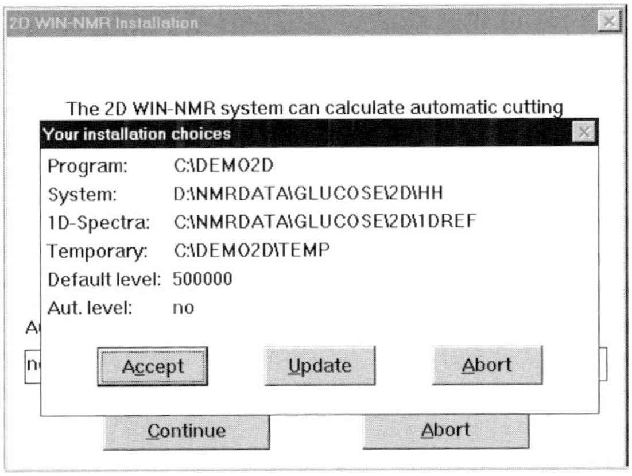

Fig. 2.2: Overview over the setup choices with 2D WIN-NMR. Note that the optional disk D: has been chosen for the System and the 1D Spectra directory.

2.3.4 Installation of GETFILE

In the following instructions describing the installation of GETFILE Demo it is still assumed that this software module is stored in a subdirectory E:\NMRPROG\GETFILE on your CD-ROM and that you want to install it in a subdirectory on disk C: of your PC. Note, however, that the full version may also be installed on other disks (e.g. D:). GETFILE will be used to demonstrate the conversion and importation of various NMR data formats. GETFILE should also be used if you plan to convert NMR data measured on older Bruker spectrometers, or measured on spectrometers of other manufacturers, or if your PC will be connected to a local network and you want to transfer and import NMR data stored on remote computers. If you want to import data using Ethernet, or the Bruker network software NMRLINK, Fastran or Kermit the appropriate software must be installed before the GETFILE program is installed. The corresponding informations will be stored in the GETFILE.INI file.

Check it in WINDOWS:

Follow again the standard procedure to install a program in the WINDOWS environment. Select the set-up program for GETFILE Demo, E:\NMRPROG\GETFILE\SETUP.EXE, on the CD-ROM and initiate the set-up process. After a few seconds the GETFILE Setup window will appear on the screen. Click the **Continue** button to proceed with the installation of GETFILE which will open a series of dialog boxes one after the other. Select the **Single User Installation** option. Confirm the proposed directory shown in the GETFILE Installation dialog box. You do not have to specify any of the transfer programs shown in this dialog box (Fig. 2.3). Consult the Bruker GETFILE manual [2.3] for further details on using GETFILE for your future work. Click on the **Continue** button to finally initialise the installation process. After successful installation a corresponding message appears on screen and the new icon for the GETFILE Demo program appears in the Program Manager WIN-NMR group. You can run GETFILE Demo from Program Manager by double clicking on this icon.

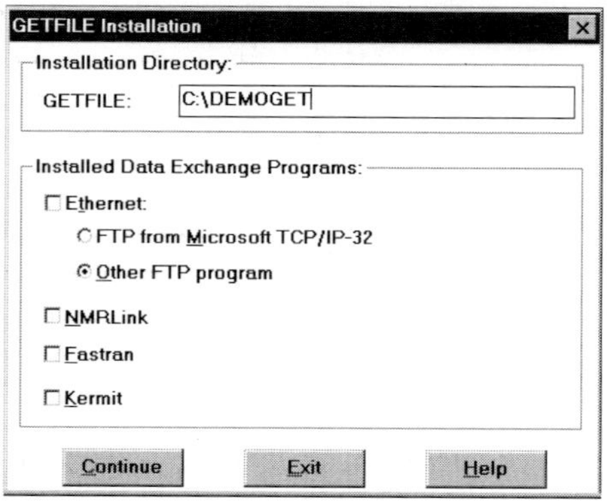

Fig. 2.3: GETFILE installation dialog box

2.3.5 Starting GETFILE, 1D WIN-NMR and 2D WIN-NMR

After successful installation, you can run the GETFILE, 1D WIN-NMR, or 2D WIN-NMR programs most conveniently by double clicking the corresponding icons in the WIN-NMR Program Manager group. This will open the 1D-WIN-NMR, 2D WIN-NMR or GETFILE application windows (Figs. 2.4, 2.5, 2.6).

Check it in GETFILE, 1D WINNMR, 2D WINNMR:

Open the GETFILE, the 1D WIN-NMR and the 2D WIN-NMR application windows (Figs. 2.4, 2.5, 2.6) by double clicking on the corresponding icon in the WIN-NMR Program Manager group.

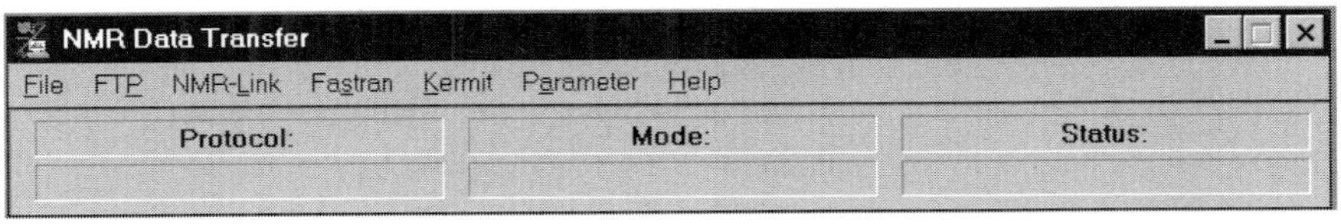

Fig. 2.4: GETFILE application window after program start

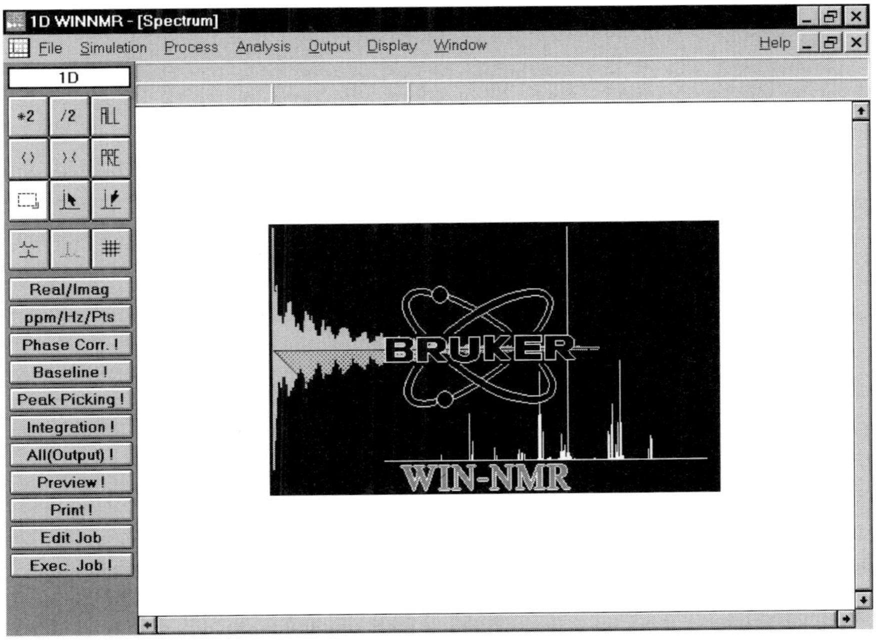

Fig. 2.5: 1D WIN-NMR application window after program start

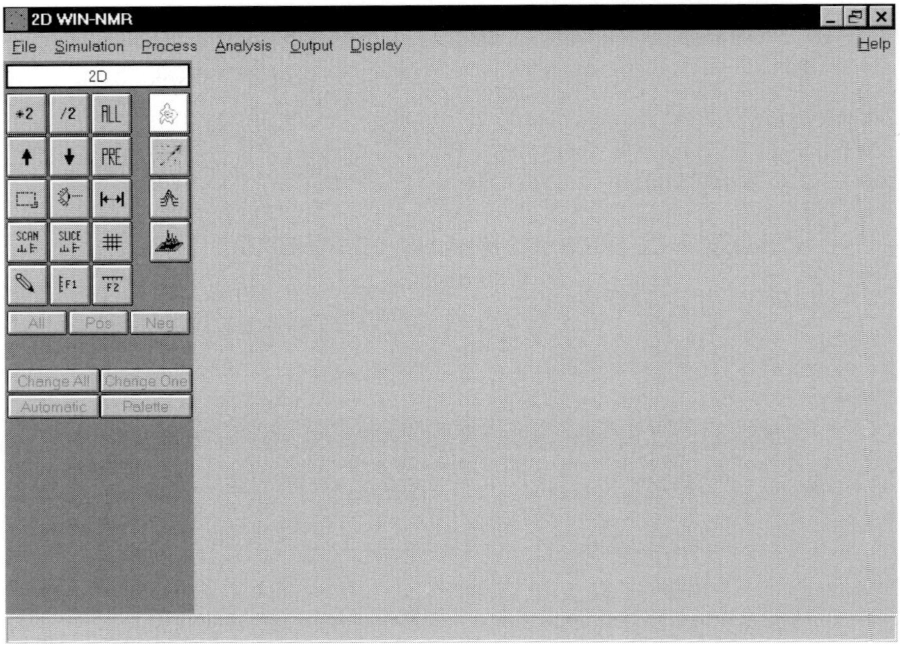

Fig. 2.6: 2D WIN-NMR application window after program start

2.4 Software- and Hardwareproblems

Although the WIN-NMR software is well tested, problems may occur, mainly caused by external disturbances or by some kind of overcharge of your PC. First check that not too many other jobs are running in the background when using any of the WIN-NMR programs. Inspect your momentary disk storage capacity and clean (see section 2.5.6) your temporary directories D:\DEMO1D\TMP and D:\TEMP if necessary. Second and if you still have problems with any of the WIN-NMR programs then either restart the corresponding WIN-NMR program(s) from your disk, or even reinstall it from your CD-ROM if necessary as decribed before.

Problems may also arise with your plotting device. In such a case check the actual printer configuration either via the corresponding WINDOWS or from within 1D WIN-NMR or 2D WIN-NMR (see chapter 4). Make sure that the settings with respect to paper size, memory capacity and orientation are correct and if there is still a problem reduce the resolution of your plotting device (e.g. 600 dpi ⇨300 dpi) or reduce the printing quality (e.g. high ⇨ medium).

2.5 NMR Data

2.5.1 Samples

Together with the WIN-NMR software a comprehensive NMR data base is delivered. The data base contains the experimental NMR data of two peracetylated carbohydrate compounds obtained from a series of 1D and 2D NMR experiments.

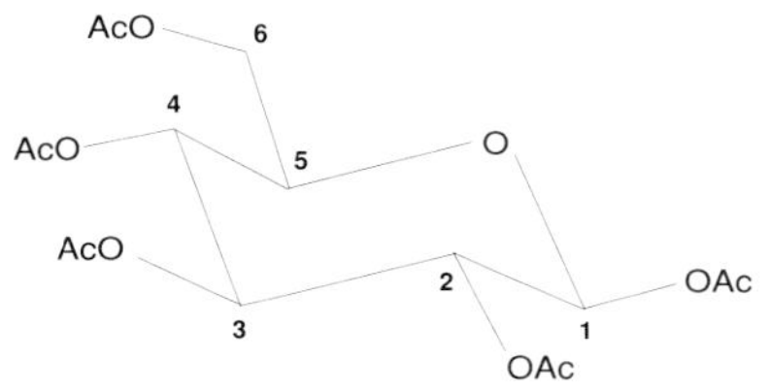

?

peracetylated β-D-glucose   peracetylated oligosaccharide

The spectra of the peracetylated β-D-glucose are used as a reference in two ways. Firstly they serve in a comparative way to let you verify your results when studying the effect of different data manipulations and the influence of different processing parameters on the processing of the experimental raw data (FID). Secondly they serve as a reference of various NMR parameters (shifts, coupling constants, ...) and give you valuable spectral information to help elucidate the unknown structure of the peracetylated oligosaccharide.

For the peracetylated β-D-glucose , raw and processed data is available, whereas for the oligosaccharide only the raw data for the same experiments is stored in the NMR data base on your CD- ROM.

The spectra for the oligosaccharide are more complex than the corresponding glucose spectra, nevertheless the raw data of the oligosaccharide may be processed in an analogous manner. Processing the oligosaccharide data gives you the opportunity to improve your skills using a more demanding case. It is the aim to process, prepare and store the NMR spectra of this unknown compound in such a way, as to have easy access to the structural information in the subsequent steps of the NMR analysis. Determining the unknown structure of this carbohydrate compound should encourage you to go further and to complete the NMR analysis with the aid of the many useful tools presented in *Modern Spectral Analysis* – volume 3 and *Intelligent Data Management* – volume 4 and to learn more about how the corresponding experiments operate as outlined in *Data Acquisition* – volume 2.

The reasons for choosing these carbohydrate compounds may be summarised as follows:

1. Both compounds display high quality 1D spectra characterised by narrow linewidths and attractive 2D spectra. In any teaching exercise, it is important that the final result is aesthetically pleasing.
2. The ^{1}H spectra of both compounds show all facets of spectral appearance, well resolved spectral regions as well as regions with heavy signal overlap and simple first order multiplets as well as complex multiplet structures caused by strong coupling effects.
3. The ^{13}C spectra of both compounds include all carbon multiplicities, i.e. CH_3, CH_2, CH and $C_{q,}$ and are ideal to demonstrate methods for spectra editing. Furthermore they include carbon nuclei with rather different T_1 relaxation times.

2.5.2 Experiments

A series of 1D and 2D NMR experiments have been performed and the corresponding data will be at your disposal on your CD-ROM. If you are not familiar with these experiments you are referred to chapter 3 where they are briefly described with their field of application, their advantages and limitations. The selected NMR experiments represent to-days most popular and successful pulse sequences for structure elucidation. This selection includes:

- The simple one pulse as well as more sophisticated multiple pulse 1D experiments for the measurement of the corresponding ^{1}H or ^{13}C spectra.
- Non-selective 1D experiments as well as selective 1D experiments including either selective weak pulses (1D ROESY, 1D TOCSY) or selective continuous radiofrequency irradiation (1D homonuclear decoupling, 1D NOE).
- Homonuclear as well as heteronuclear 2D shift correlation experiments ($^{1}H/^{1}H$-COSY, $^{1}H/^{13}C$-COSY), involving the perturbation of either one or two types of nuclei respectively and in the heteronuclear case including both the conventional, direct ^{13}C detection, as well as the more sensitive, indirect ("inverse" or "reverse") ^{1}H-detection.
- The ^{1}H J-resolved 2D experiment.

The second goal was to introduce the newcomer in this field to those modern experiments, which are best suited to detect and measure the main and for the structure elucidation most useful NMR parameters. This includes experiments dedicated to determining the chemical shifts of protons and carbons, to detecting their mutual homo- and heteronuclear connectivities within a molecule via one-bond ($^{1}J_{CH}$) or multiple-bond ($^{n}J_{CH}$) scalar spin-spin interactions, determining $^{1}J_{CH}$ based carbon multiplicities, measuring homo- and heteronuclear coupling constants, and to obtaining direct as well as related relaxation parameters governed by dipolar coupling among spins, *i.e.* T_1 or T_2 and NOEs respectively. By presenting all the important NMR parameters and the corresponding NMR data the user experiences that it is this heteronuclear, multiple parameter approach, which makes high resolution NMR so successful in structure elucidation.

Last, but not least, the choice of experiments should provide the user with NMR data allowing him to apply and test almost all the processing options available in the 1D WIN-NMR and 2D WIN-NMR software modules.

For more details concerning the various pulse techniques, the corresponding spectra and the structure information which may be obtained, the reader is referred to chapter 3 (section 3.4) or *Data Acquisition* – volume 2 of this series.

2.5.3 Experimental Conditions

Both samples, the peracetylated β-D-glucose (30mg) and the peracetylated oligosaccharide of unknown structure (6mg) were dissolved in $CDCl_3$ (99.8% D), doped with a trace of tetramethylsilane (TMS) as the internal standard.

All spectra were measured on a Bruker DRX 500 spectrometer, at 500.13MHz (^{1}H) and at 125.76MHz (^{13}C). Sample spinning (20Hz) was used for all 1D experiments with the exception of the 1D NOE and 1D ROESY experiments. As is now common, all 2D experiments were performed with the sample static. All experiments were performed at ambient magnet temperature without any special temperature control.
The spectrometer was equipped as follows:

- Triple Inverse 5mm TBI probehead for ^{1}H and ^{13}C observation with one additional channel for X observation or decoupling (choice of one nucleus within the range ^{31}P – ^{109}Ag), a Z-gradient coil and operable in the temperature range –50° to +80°C. 90° pulselenghts for ^{1}H and ^{13}C were 6.9us (–6dB) and 13us (–4dB) respectively.
- For frequency selective experiments either CW-decoupling or weak shaped pulses were applied. CW power levels of 60Hz (50dB) for the 1D homonuclear decoupling experiment and 10Hz/5Hz (68dB/74dB) for the 1D NOE experiment were used. TOPHAT shaped 90° pulses of 160ms (60dB) were used for the 1D TOCSY and the 1D ROESY experiments respectively.
- For spin-lock experiments a 90° pulselength (pulsepower) of 30us (10dB) was used for the 1D/2D TOCSY experiments and a series of 180° pulses with a pulselength (pulsepower) of 124us (16dB) was applied with the 1D/2D T-ROESY experiments.
- Broadband 5mm probehead for observation of nuclei in the range from ^{15}N to ^{31}P with ^{1}H decoupling, operable in the temperature range –160° to +180°C. 90° pulselenghts for ^{1}H and ^{13}C were 10.8us (–6dB) and 8us (–4dB) respectively. This probehead was used for ^{13}C detected experiments only.
- 16 bit digitizer
- Digital filter
- Bruker Smart Magnet control System (BSMS) including the digital deuterium lock system and the BOSS1 shim system
- Gradient spectroscopy accessory BGPA10 with a maximum gradient strength of 30G/cm. Gradient pulses of 1.5ms with a recovery time of 150us were applied in the appropriate experiments.
- The spectrometer was connected to an INDY R4600 (133MHz) work station equipped with a 2 GB hard disk and running under the XWIN-NMR Software (Version 1.0).

For more experimental details please inspect the parameter files of the corresponding experiments accessible from within 1D and 2D WIN-NMR.

2.5.4 Directory Structure

Fig. 2.7 shows the main directory structure and the structure of two subdirectories of the NMR data base stored on your CD-ROM as displayed by the WINDOWS File manager. If it is planned to copy the NMR data from your CD-ROM onto your hard disk to speed up further processing, the same directory structure should be established there (see 2.5.5). Note, however, that 1D WIN-NMR generates additional auxiliary subdirectories (AU, DAT, NMR, PC, TMP) in the directory DEMO1D on your harddisk.

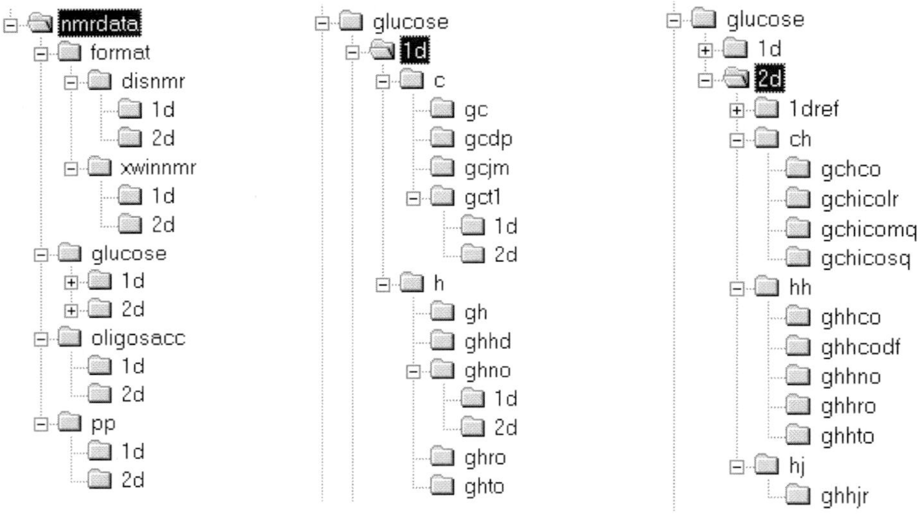

Fig. 2.7: Main directory structure of the NMR data base NMRDATA (left), of the GLUCOSE\1D (middle) and of the GLUCOSE\2D subdirectory (right).

The main directory NMRDATA consists of the four subdirectories FORMAT, GLUCOSE, OLIGOSAC and PP. The subdirectories GLUCOSE and OLIGOSAC include the data of the various NMR experiments applied to the peracetylated β-D-glucose and the peracetylated oligosaccharide of unknown structure respectively. The subdirectory FORMAT includes two further directories DISNMR and XWINNMR; these additional directories include non-converted 1D and 2D data measured and directly imported from Bruker spectrometers running under DISNMR and XWIN-NMR or UXNMR/XWINNMR respectively. The data contained in the FORMAT directory is used to demonstrate the different structures of non-converted data files and to illustrate the conversion of such data files into the WINNMR format. Any experimental NMR data must first be transferred into the WINNMR format to be processed with the WIN-NMR software tools. PP contains the pulse programs used to measure the NMR data stored in the directories GLUCOSE and OLIGOSAC.

The subdirectories GLUCOSE and OLIGOSAC have both the same substructures:

- The directory 1D consists of the subdirectories C and D containing the data sets for a series of 1D ^{1}H and ^{13}C experiments. The data for two experiments (1D NOE, T_1 Inversion Recovery) is stored in two different ways, since these kind of experiments are performed in a pseudo-2D mode on the Bruker DRX spectrometer, yielding 2D data matrices. The data is therefore stored in different subdirectories either in its original 2D format (as a SER-file) in the subdirectory 2D (e.g. D:\NMRDATA\GLUCOSE\1D\C\GCT1\2D or as a series of extracted 1D data files in the subdirectory 1D (e.g. D:\NMRDATA\GLUCOSE\1D\C\GCT1\1D. For the frequency selective 1D ROE and the 1D TOCSY experiments data obtained after selectively perturbing selected spins are stored in subdirectories FREQ (e.g. D:\NMRDATA\GLUCOSE\1D\H\GHRO\FREQ) and data acquired using different mixing times (see chapters 3.3.1.3 and 3.3.1.5) are stored in subdirectories TMIX (e.g. D:\NMRDATA\GLUCOSE\1D\H\GHTO\TMIX). The two subdirectories FREQ and TMIX are not shown in Fig. 2.7.
- 2D is composed of three subdirectories. 1DREF contains 1D ^{1}H and ^{13}C spectra measured with high digital resolution, used as projections for the 2D spectra, while CH, HH and HJ contain the data for heteronuclear $^{1}H/^{13}C$- and homonuclear $^{1}H/^{1}H$-2D shift correlation and for ^{1}H-J (coupling) resolved experiments respectively.

The file names, e.g. GHHCODF, of the NMR data sets as shown in Fig. 2.7 and stored in the subdirectories C, H, 1DREF, CH and HH include information on the sample, the type of spectrum (1D/2D) and the experiment itself. The meanings of the various abbreviations are listed below:

sample:

G peracetylated β-D-glucose

O oligosaccharide

nuclei and type (1D, 2D) of spectrum:

H 1D ^{1}H spectrum

C 1D ^{13}C spectrum

1DREF 1D ^{1}H and ^{13}C spectra (1D projection spectra for 2D spectra)

HH 2D $^{1}H/^{1}H$ shift correlation spectrum

HJ 2D 1H-J (coupling) resolved spectrum

CH 2D $^{1}H/^{13}C$ shift correlation spectrum with ^{13}C detection

CHI 2D $^{1}H/^{13}C$ shift correlation spectrum with ^{1}H detection (“inverse” mode)

type of experiment:		**example:**
_	basic 1D experiment	GH, GC
CO	COSY experiment	GHHCO
	CH-COSY experiment ($^{1}J_{CH}$)	GCHCO
CODF	DQ-filtered COSY experiment	GHHCODF

COLR	CH-COSY experiment ($^{n}J_{CH}$)	GCHICOLR
COMQ	CH-COSY experiment ($^{1}J_{CH}$, multiple quantum evolution)	GCHICOMQ
COSQ	CH-COSY experiment ($^{1}J_{CH}$, ingle quantum evolution)	GCHICOSQ
DP	DEPT experiment	GCDP
HD	experiment with homodecoupling	GHHD
JM	J-modulated (APT) experiment	GCJM
JR	J-resolved experiment	GHHJR
NO	NOE experiment	GHNO
	NOESY experiment	GHHNO
RO	ROE experiment	GHRO
	ROESY experiment	GHHRO
TO	TOCSY experiment	GHTO,GHHTO
T1	T_1 experiment	GCT1

In chapter 3 the basic theory and the application of these experiments to obtain structural information is briefly discussed and examples of corresponding spectra obtained for the peracetylated β-D-glucose are depicted.

2.5.5 Copying the NMR Data from the CD to your Hard Disk

In the following instructions describing how to copy the NMR data base it is assumed that this data base is stored in a subdirectory E:\NMRDATA of your CD-ROM and that you want to copy it to disk D: of your PC as already prepared with the installation of 1D and 2D WIN-NMR (see 2.3.2, 2.3.3). Installation on other disks (e.g. C:), however, is also possible. In principle you have also direct access to the NMR data on your CD-ROM, but for several reasons it is recommended to copy the whole data base to one of the harddisks of your PC.

Check it in WINDOWS:

Start the WINDOWS file manager (WINDOWS Explorer) and check there is 150-200 MBytes of space available on one of your hard disk(s). Follow one of the standard procedures for copying files from one to another directory. Consult the WINDOWS manual, or first study the *Check it* instructions for copying a file or directory given in the next section (2.5.6), if necessary. Select in your CD-ROM directory the subdirectory NMRDATA, and copy it to disk D: on your PC. The directory NMRDATA\GLUCOSE\1D\H and a few auxiliary directories therein (see 2.5.4) have already been created during installation of 1D WIN-NMR. Answer the corresponding question for overwriting all files stored under the same name with yes. Use the WINDOWS file manager (WINDOWS Explorer) to inspect the newly created NMR data directory NMRDATA on your hard disk. Click on several items in this directory and familiarise yourself with the directory structure.

2.5.6 Useful Options in the MS WINDOWS 95 Operating System

If you have not yet used the MS WINDOWS 95 operating system, study first the corresponding manual and try the subsequent *Check its* on your PC to become familiar with a few options which will be most useful when working with 1D WIN-NMR, 2D WIN-NMR and the NMR data base. The **WINDOWS Explorer** (WINDOWS file manager) dedicated to manage your directories and programs may be used either outside of 1D/2D WIN-NMR, or may be started from within one of these programs. Its "drag and drop" functionality (see below) can be put to good use to most conveniently read in the selected data file and to display it in the corresponding WIN-NMR windows (see chapter 4). Very similar options are available with the WINDOWS 3.1 and 3.11 versions. Please consult the corresponding introductory WINDOWS manuals if necessary.

Check it in WINDOWS 95: Create a new directory

Consult your WINDOWS 95 manual and study the section for creating a new directory or folder. Double click **My Computer** and then double-click the disk drive or directory D:\NMRDATA in which you want to place the new folder. On the **File** menu point to **New** and then click **Folder**. Type the name of the new folders TEST1, TEST2 and TEST3 and press the **Enter** key. Check the three newly created subdirectories of D:\NMRDATA.

Check it in WINDOWS 95: Copy or move a file or a directory

Consult your WINDOWS 95 manual and study the sections for copying or moving files and directories. Note that several possibilities exist for these purposes. Double-click **My Computer**, find the file HDIS.001 in the directory D:\NMRDATA\FORMAT\DISNMR\1D\H you want to copy and click it. Click **Edit** which will open a corresponding pull-down menu and then click **Copy** to copy the selected file. Open now the directory D:\NMRDATA\TEST1 where you want to place the file, click **Edit** and then **Paste**. Select now the file CDIS.001 in the directory D:\NMRDATA\FORMAT\DISNMR\1D\C and copy it in the same way into the destination directory D:\NMRDATA\TEST2. Check the new entries in the TEST1 and TEST2 subdirectories.

Proceed as before, select and click the file HDIS.001 in the directory D:\NMRDATA\TEST1 you want to move and click it. Click **Edit** and then **Cut** to move the file. Open now the directory D:\NMRDATA\TEST2 where you want to place the file, click **Edit** and then **Paste**. Check the new entry in the TEST2 subdirectory.

Alternatives: Two much simpler ways of using the right mouse button to copy or move a file exist:

Double-click **My Computer**, find and click the file CDIS.001 in the directory D:\NMRDATA\TEST2 you want to copy or move with your right mouse button. To copy this file, click **Copy,** open the destination directory D:\NMRDATA\

TEST1 where you want ot place the selected file and use the right mouse button to click an empty part of the selected window and click **Paste**.

To move the file HDIS.001001 from the directory D:\NMRDATA\TEST2 to the directory D:\NMRDATA\TEST1 proceed as above but click **Cut** instead of **Copy**. Check the entries in the two directories.

Enter the WINDOWS Explorer to exploit a second alternative for copying and moving files. Open the directory D:\NMRDATA\TEST1 and click with your right mouse button on the file HDIS.001. Hold down this mouse button while you move the symbol of the selected file with your mouse from its original place to the destination directory D:\NMRDATA\TEST2 ("drag and drop" method). Click **Copy Here** to copy the file. Select the CDIS.001 file in the directory D:\NMRDATA\TEST1, proceed as above but click **Move Here** to move the file. Since this file already exist in this directory a corresponding message for replacement appears.

Series or Files: To copy or move series of files select the files to be copied or moved in the same way as above one after the other, while pressing and holding down at the same time the **CTRL** key. Release this key and use the "drag and drop" method as described above to copy or move the whole series of selected files to the destination directory. Select the two files HDIS.001 und CDIS.001 in the directory D:\NMRDATA\TEST2 and copy them into the directory D:\NMRDATA\TEST3 using the option for selecting and copying series of files as described above. Check the new entries in the TEST3 subdirectory.

Check it in WINDOWS 95: Copy Files to a Floppy Disk

When you want to copy NMR data files from your hard disk to a floppy disk the easiest way is to use **My Computer**. Double-click **My Computer**, find the file you want to copy, then click on it. On the **File** menu point to **Send To** and then click the drive where you want to copy the file or directory. You can again select multiple items by pressing and holding down the **CTRL** key as you click each item you want.

Check it in WINDOWS 95: Show the properties of a file

Double-click **My Computer**, find the file HDIS.001 in the directory D:\NMRDATA\TEST2 and click **Properties**. Several properties such as the file's size, its directory or its attributes are displayed in a menu and may be inspected or changed.

Check it in WINDOWS 95: Rename a file or directory

Double-click **My Computer**, find the file CDIS.001 in the directory D:\NMRDATA\TEST2 and click **Rename**. Use DISNMR.001 as the new name.

Check it in WINDOWS 95: Search for files

Consult your WINDOWS 95 manual and study the section for finding something on your computer. Click the **Start** button and then point to **Find**. Click **Files or Folders...** and then click the **Named** box. Then type the name of the file or directory you want to find. As an example enter <*.SER> as the search mask (SER means 2D raw data file) and in the **Look in** box enter <D:\NMRDATA> as the directory where to search. To start the search click **Find Now**.

Check it in WINDOWS 95: Delete a file or directory

Consult your WINDOWS 95 manual and study the section for deleting files or directories. Double-click **My Computer** and find the file D:\NMRDATA\TEST2\1D\HDIS.001 you want to delete, and then click it. On the File menu click **Delete**.

Select the subdirectories TEST1, TEST2 and TEST3 and again apply the **Delete** command to remove all the remaining files. Note that for deleting series of files the files are selected in the same way one after the other as before, while pressing and holding down at the same time the **CTRL** key. Release this key and apply the **Delete** command. Delete the three subdirectories TEST1, TEST2 and TEST3 as well.

2.6 Data Formats

Bruker uses two different formats (UXNMR/XWINNMR, DISNMR) for the original NMR data depending on the type of spectrometer and the type of data system. Newer spectrometers (DMX, DRX, DSX, AMX) connected to an INDY workstation, an ASPECT Station 1 or an ASPECT X32 computer, all running under the UNIX operating system and using the UXNMR, or most recently the XWIN-NMR software, yield NMR data in the UNIX format. Older systems (AM, AC, MSL) connected to an ASPECT 2000 or ASPECT 3000 computer running under the ADAKOS operating system and using the DISNMR software use a different format.

Data in either format must first be converted into a form that WIN-NMR understands before any processing steps may be performed with either 1D WIN-NMR or 2D WIN-NMR. 1D WIN-NMR and GETFILE include tools for converting both data formats into this WINNMR format. 2D WIN-NMR converts 2D data files in the UXNMR/XWINNMR format – but not in the DISNMR format – automatically (see section 2.7.2).

Conversion into the WINNMR format may be done for original data files stored on your disk or may be coupled with the transfer of data from remote computers (i.e. the spectrometer itself or an intermediate file-server) as described in section 2.7.

2.6.1 WINNMR Format

1D WIN-NMR and 2D WIN-NMR use the WIN-NMR data format. With the exception of the NMR data stored in the directory D:\NMRDATA\FORMAT, all other NMR data stored in the directories D:\NMRDATA\GLUCOSE and D:\NMRDATA\OLIGOSAC supplied on CD-ROM is in the WINNMR format.

When NMR data is converted into the WINNMR format it is converted into a number of files stored in a directory called <NAME>. For 1D- and 2D WIN-NMR, the directory <NAME> has the following structure:

```
1D WIN-NMR                      2D WIN-NMR
              E   P                           E   P
            ┌───┐┌───┐                      ┌───┐┌───┐
<NAME>\     0 0 1 0 0 1.AQS     <NAME>\     0 0 1 0 0 1.FA1
            0 0 1 0 0 1.FQS                 0 0 1 0 0 1.FA2
            0 0 1 0 0 1.FID                 0 0 1 0 0 1.FP1
            0 0 1 0 0 1.1R                  0 0 1 0 0 1.FP2
            0 0 1 0 0 1.1I                  0 0 1 0 0 1.SER
                                            0 0 1 0 0 1.RR
                                            0 0 1 0 0 1.II
                                            0 0 1 0 0 1.IR
                                            0 0 1 0 0 1.RI
```

E and **P** denote the experimental and the processing number of the data file respectively. For a given sample the raw data of up to 999 different experiments may be stored under the same data file name. The data of each for these experiments can be processed in several ways and up to 999 different results may be stored separately together with the corresponding processing parameters. The meaning of the various file name extensions is shown below:

	1D Data Set	2D Data Set
Acquisition Parameters	*.AQS	*.FA1 AND *.FA2
Processing Parameters	*.FQS	*.FP1 AND *.FP2
Raw Data File	*.FID	*.SER
Processed Data – Real	*.1R	*.RR
Processed Data – Imaginary	*.1I	*.II, *.IR, AND*.RI

Note: For peracetylated β-D-glucose both the raw data (FID, SER) and the corresponding spectra (1R, 1I, RR, ...) are stored on the CD-ROM. The fully processed spectra serve as a reference and are stored under the processing number 999 (e.g. D:\NMRDATA\GLUCOSE\1D\H\GH\001999.1R, D:\NMRDATA\GLUCOSE\2D\HH\GHCO\001999.RR). This allows you to process the raw data according to your own ideas, to store it under the processing number 001-998 and to compare your processing result with the corresponding reference spectrum (999).

Note: ∗.IR and ∗.RI are only present if the spectrum and the corresponding experiment is phase sensitive (see section 3.4). Additional files such as ∗.PLT, ∗.TIT, ∗.TI2 etc. may appear with your NMR data and contain additional information concerning the plotting parameters, the spectrum title etc.

Check it in WINDOWS:

Using WINDOWS Explorer select one of the 1D NMR data directories stored in the directory D:\NMRDATA\GLUCOSE\1D and inspect the WINNMR data structure using the WINDOWS file manager (WINDOWS Explorer). Similarly inspect the data structure of one of the 2D NMR data files stored in the directory D:\NMRDATA\GLUCOSE\2D. Check the processing number of the corresponding spectra.

Check it in 1D WIN-NMR:

Start the 1D WIN-NMR Demo program by clicking on the corresponding icon. From the **File** pull-down menu choose the **Open** option and select in the dialog box the directory D:\NMRDATA\GLUCOSE\1D\H\GH. Check the WIN-NMR symbol **W** (in blue)) at the beginning of each entry in the file list of the selected subdirectory. This directory should contain FIDs (∗.FID) and spectra (∗.1R). Check the processing number of the corresponding spectra. Use the **Cancel** command to leave this dialog box.

Check it in 2D WIN-NMR:

Start the 2D WIN-NMR program by clicking on the corresponding icon. From the **File** pull-down menu choose the **Open** option and select in the dialog box the directory D:\NMRDATA\GLUCOSE\2D\HH\GHHCO. Check the WIN-NMR symbol **W** (in blue) at the beginning of each entry in the file list of the selected subdirectory. This directory should contain 2D raw data (∗.SER) and spectra (∗.RR) files. Use the **Cancel** command to leave this dialog box.

2.6.2 UXNMR/XWINNMR Format

NMR data measured on a modern Bruker NMR spectrometer (DMX, DRX, ARX) is stored as a directory in the UXNMR/XWINNMR format. Under the UNIX operating system, data files stored as <NAME> are themselves directories.

For 1D and 2D UNIX data files, the corresponding directories <NAME> have the following structures shown below. **E** and **P** again denote the experimental and the processing number of the data file respectively. For a given sample the raw data of up to 999 different experiments may be stored under the same data file name. The data for each of these experiments can be processed in several ways and up to 999 different results may be stored separately together with the corresponding processing parameters.

1D Data

	E		P
<NAME>\1		\PDATA	\1
	ACQU		PROC
	ACQUS		PROCS
	FID		1R
			1I

2D Data

	E		P
<NAME>\1		\PDATA	\1
	ACQU		PROC
	ACQUS		PROCS
	ACQU2		PROC2
	ACQU2S		PROC2S
	SER		2RR
			2II
			2IR
			2RI

The parameter files ACQU* and PROC* contain acquisition and processing parameters, and the files FID and SER contain the acquired 1D or 2D raw data, respectively. For data measured in quadrature detection mode (see volume 2 of this series), the data points acquired with channel A and B alternate within an FID. A 1D FID file contains a single FID with TD(F2) points; the time domain size TD (see chapter 5) is stored in the acquisition status parameter file ACQUS. A 2D SER file contains TD(F1) FIDs each with TD(F2) points. TD(F1) is the parameter TD in the file ACQU2S and TD(F2) is the parameter TD in the file ACQUS. The files PROC, PROCS, PROC2 and PROC2S contain processing parameters as set by the operator on the spectrometer. They are not obligatory and may be altered and adjusted to best fit the users needs. Processed 1D data is stored in two files 1R and 1I, corresponding to the real and imaginary part of a 1D spectrum. Like for 1D FIDs, the data points are stored as a sequence of 32 bit integers. 2D processed data is also stored as 32 bit integers. The real part of the spectrum is contained in the file 2RR and the imaginary and mixed parts in the files 2II, 2RI and 2IR. The latter two are only present if the spectrum is phase sensitive. All files are stored in the so-called submatrix format. The submatrix dimensions are given by the status parameters contained in the files PROCS and PROC2S.

Additional files, FORMAT.TEMP, PULSEPROGRAM, VDLIST, TITLE, OUTD, PARAM.TXT, META and others may also be present if you have imported your NMR data directly from a Bruker spectrometer. These files contain additional information and settings initialized by the spectrometer operator and relate to the acquisition pulse program, lists of variable delays, spectrum title, the spectral layout and others and are non-essential for off-line data processing.

Important: The files ACQU, ACQUS, FID, PROC and PROCS are essential for proper processing of a 1D data file and must be stored on your hard disk before you start any processing using WIN-NMR. Processing of a 2D NMR data file requires the files ACQU, ACQUS, ACQU2, ACQUS2, SER, PROC, PROCS, PROC2 and PROCS2 to be available on your hard disk. If any of these files are missing an error message will appear and no processing will be performed. If transformed spectra are stored on disk then both the real and imaginary files must be stored to allow the readjustment of the spectrum phase (see sections 5.2.3 and 5.2.5). For a 1D spectrum this is *.1R and *.1I respectively. For a phase sensitive 2D spectrum (see chapter 3) this is *.RR, *.II, *.IR and *.RI. If any of these files are missing, the data will have to be processed again.

Check it in WINDOWS:

Select the 1D and the 2D NMR data files HUX and HHUX in the directory D:\NMRDATA\FORMAT\XWINNMR\1D\H and D:\NMRDATA\FORMAT\XWIN NMR\2D\HH respectively and inspect their data structures using the WINDOWS file manager (WINDOWS Explorer). Check the differences between a 1D and a 2D data file.

Check it in 1D WIN-NMR:

Start the 1D WIN-NMR program by clicking on the corresponding icon. From the **File** pull-down menu choose the **Open** option and select in the dialog box the directory D:\NMRDATA\FORMAT\XWINNMR\1D\H\HUX. Check the UNIX symbol **UX** (in red) in front of the filenames in the selected subdirectory. Do not click on any of the data files in this moment, but use the **Cancel** key to leave this dialog box.

2.6.3 DISNMR Format

1D DISNMR data files are simply stored as <NAME.nnn>, with nnn being the experiment number (001 - 999). 2D DISNMR data files are stored as <NAME>.SER. No further file structure is recognized.

In principle there is no label, if not set by the user, to identify the given DISNMR file as an FID or spectrum, or if the NMR data is from a ^{1}H or ^{13}C experiment. However the Bruker automation software, primarily developed to connect older type spectrometers (AC, AM) to a sample changer, allows the user to structure the name in such a way that it carries additional information with respect to the type of experiment. The user is referred to the corresponding automation manual available from Bruker and to the name conventions set by the key NMR operator at your site.

Check it in WINDOWS:

Select the directories D:\NMRDATA\FORMAT\DISNMR\1D\H, D:\NMRDATA\ FORMAT\DISNMR\2D\HH and D:\NMRDATA\FORMAT\ DISNMR\2D\CH and inspect the structures of the NMR data files using the WINDOWS file manager (WINDOWS Explorer).

Check it in 1D WIN-NMR:

Start the 1D WIN-NMR program by clicking on the corresponding icon. From the **File** pull-down menu choose the **Open** option and select in the dialog box the directory D:\NMRDATA\FORMAT\DISNMR\1D\H. Check the **DIS** (in red) symbol in front of the filename in the selected subdirectory. Do not click on this file in this moment, but use the **Cancel** key to leave this dialog box. Make sure that the DISNMR file type is crossed.

2.6.4 NMR Data Formats of other Manufacturers: Varian, JEOL, GE

In many cases the NMR fascilities in universities and industry are equipped with spectrometers manufactured by companies other than Bruker/Spectrospin. The NMR data from Varian (Gemini and Unity), JEOL (Alpha, GX, EX and Lambda) and GE spectrometers may also be processed by 1D WIN-NMR and 2D WIN-NMR provided that the data is first converted into WIN-NMR data format. The data should be transferred to a PC, preferably via an internal network, and then the appropriate routine in the GETFILE module is used for the conversion. To use this conversion tools a copy protection dongle (WIBU-key) – not delivered with this education package – must be installed on your PC. If you plan to install this option, please refer to the instructions given in the GETFILE manual [2.3] and contact your Bruker/Spectrospin representative.

Check it in GETFILE:

Start GETFILE and inspect the various dialog boxes available in the **File** pull-down menu for conversion of NMR data formats of other manufacturers into the WINNMR format.

2.6.5 Other Formats: ASCII, JCAMP-DX

Your data files (FIDs, spectra) are usually stored as binary (WINNMR format) on your PC but they may also be converted into and stored as ASCII or JCAMP-DX format files.ASCII format stores the currently displayed spectrum together with the most important parameters as a single file using ASCII characters. Several options (whole/displayed part of spectrum/FID, with/without header, real/complex data) may be defined according to your needs.JCAMP-DX format stores the data points of the whole spectrum together with all acquisition and processing parameters in a single ASCII file using a specific protocol defined as JCAMP-DX5 format. Several options (FID/real spectrum/complex spectrum, four types of compression) are available and may be set according to your needs. This format is useful for exchanging spectroscopic data with other hardware platforms where it may be processed with NMR software tools other than the WIN-NMR modules.

Such data formats (ASCII, JCAMP-DX) can be read by any text editor and are a suitable format for transferring files in a compressed form via international networks, e.g. the INTERNET using e-mail. For this purpose load the file you want to export from within 1D WIN-NMR, use the **Save as...** option, select in the corresponding dialog box the JCAMP-DX format and hit the **OK** button. Select in the next dialog box (**JCAMP-DX Option**) the **Diff/Dup** compress mode and hit the **OK** button in this second dialog box, which will start the conversion and will change the files's extension to DX.. The converted file may then be incorporated as an attachment into an e-mail file and may be exported via Internet. If you receive such an e-mail, first cut the header, including mail informations and eventually added text at the end of the mail and store the residual data file under a name with the extension DX, e.g. D:\EMAILNMR\001001.DX. If you load this file from within 1D WIN-NMR it will be automatically converted to the WIN-NMR format and will be displayed on the screen for further processing.

An even more convenient alternative to export/import NMR data using Internet will be discussed in the next section.

Additional data formats (metafiles) to export spectra, parameters or titles into word processing or desk top publishing packages are discussed in chapter 4.

2.7 Data Import and Export

If you plan to use the full version WIN-NMR software tools to process your NMR data you will have to import your NMR data files onto your PC and if you want to store your processed data on remote computers or if you want to submit your NMR data to other users you will have to export your data. The most basic, but archaic method for data import/export, would be to use high density floppy diskettes. The main disadvantage of floppy diskettes are their long copy times and small storage capacity. As an example a standard 1.44 MB diskette will store one single 2D data matrix of size F1xF2 = 512x512 or 1024x256 or 2048x128, ... The corresponding digital resolutions are in most cases not high enough to be of practical use (although this will depend on the spectrometer's operating frequency). It is unlikely that the storage capacity of floppy diskettes will increase substantially in the near future, so with the exception of 1D data sets, using this method for file import cannot be recommended. The use of local networks is certainly the simplest and most popular way to import NMR data measured and stored on remote computers; copying is much faster and there is almost no limit regarding file size subject, of course, to available disk space on your PC. However additional hardware and software must be installed to connect your PC to a local network. When the necessary hardware and software requirements have been met, data can be read directly into the 1D WIN-NMR and 2D WIN-NMR programs from the hard disks of workstations using the UNIX format such as the SGI INDY and Bruker ASPECT Station 1. On these types of work station data can be stored in either WIN-NMR, UXNMR/XWINNMR, DISNMR, DISMSL or JCAMP-DX5 formats. For data stored on computers using formats other than UXNMR/XWINNMR (e.g. DISNMR) additional conversion software and different procedures for data transfer and data conversion must be applied (see section 2.7.2.2).

Various protocols are available for importing data from a remote computer to your PC. The simplest and most elegant way to perform this task is to use either NFS (Network File Server) or FTP (File Transfer Protocol). An example of how NFS and FTP may be used within a local network, including the conversion of the corresponding data format into the WINNMR format, is outlined in section 2.7.2.

Please check with the system administrator responsible for your local network, which transfer protocols are available at your local site and what additional hardware and software will be required for networking your PC.

If you have installed MAPI (mail application interface) software on your PC, you may exploit the MS-WINDOWS mailslot-function to e-mail NMR data directly to and from your PC. The full version of 1D-WIN-NMR allows you to export/import FIDs, spectra, tables, text-files, relaxation data and metafiles to/from other users of (the full version of) 1D WIN-NMR. Both JCAMP-DX5 and Bruker specific binary format are supported. Compared to the procedure outlined in section 2.6.5 this is an even more convenient way for exporting/importing NMR data via Internet. For further details refer to the 1D WIN-NMR manual [2.1] or contact your Bruker/Spectrospin representative.

2.7.1 Network-Example

As an example the authors local NMR network (Department of Chemistry and Biochemistry, University of Berne, Switzerland) is outlined below (Fig. 2.8). It represents a network situation, typical for universities, but probably not representative for industry, with other standards and additional demands.

Two Bruker DRX-spectrometers, each connected to a INDY-workstation (INDY WS) use FTP to communicate and exchange data files via a thin-wire Ethernet with a SUN workstation, acting as a bridge to the central network (thick-wire Ethernet). FTP is used to send data from both DRX spectrometers via the bridge to the central server workstation (SERVER WS).

Two Bruker AC spectrometers are connected to a Bridge-PC using NMRLINK. These two PCs are also connected to the central network (thick-wire Ethernet) and exchange data with the central Server Workstation using FTP. A series of PCs, located in laboratories and serving as remote NMR processing stations, are connected via the central network to the central Server Workstation. Most of these remote PCs use NFS for transferring NMR data. Since this communication is restricted to the transfer of NMR data from the central Server to the PC no passwords are required.

The PCs of the system administrator and a few special users have direct access to the DRX-spectrometers via the central Server using the FTP protocol. This special group have direct access to any files in the spectrometer's data system, i.e. data files, various lists, pulse programs, ... and may transfer files to and from the spectrometers. In addition, they also have the option to create or delete directories, to modify pulse programs and to do other jobs on the spectrometer's data system. As is usual in security sensitive situations, this special group require a password to access the spectrometer.

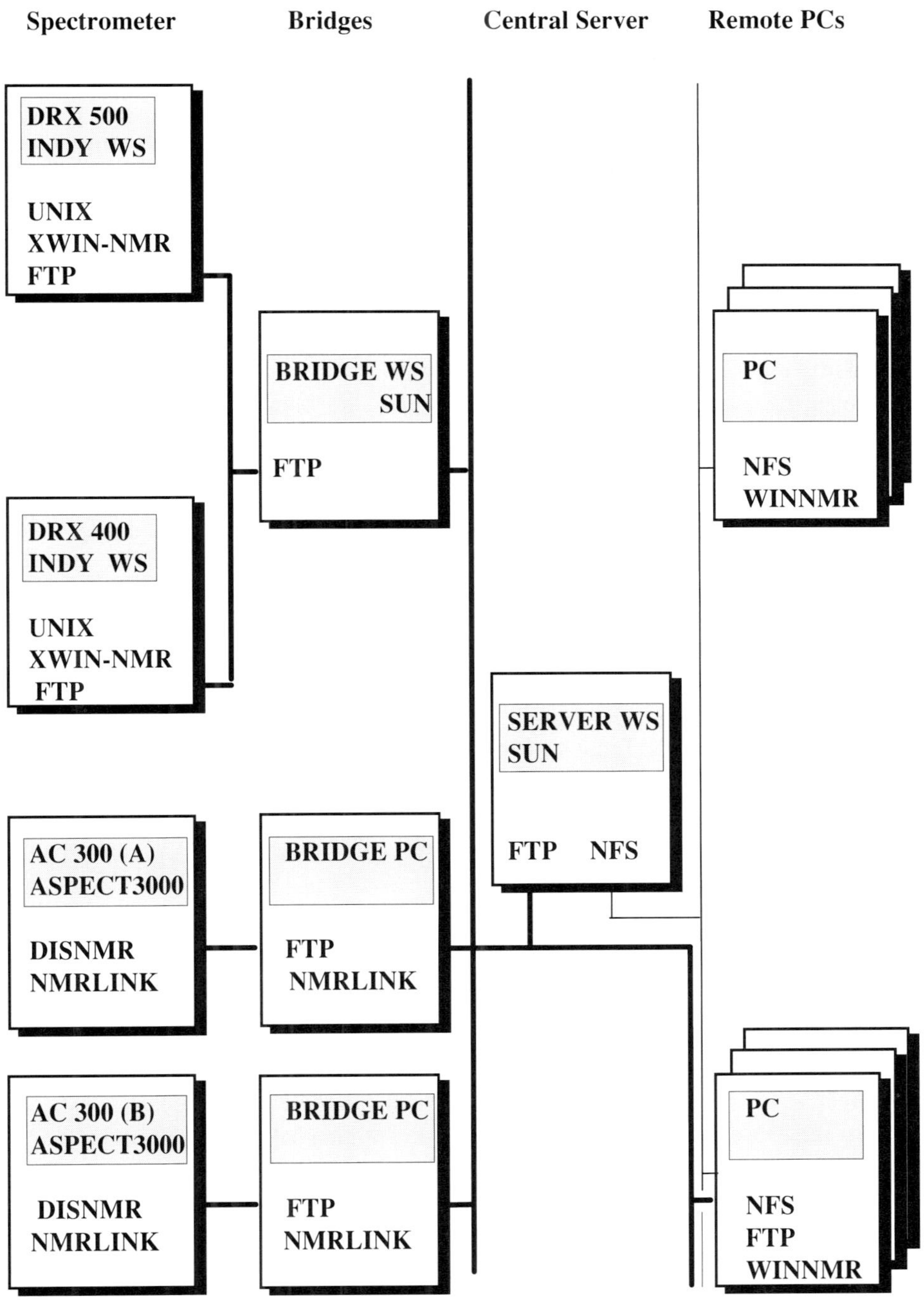

Fig. 2.8: Example of a network for remote processing of NMR data

2.7.2 Transfer and Conversion of NMR Data stored on Remote Computers

In principle any NMR data file may be transferred and stored on your PC hard disk in its original format (UXNMR/XWINNMR, DISNMR). Alternatively the NMR data file may have already been converted into the WINNMR format on the remote computer before being copied to your PC. The most usual way, however, is to import the file in its original format and to convert it into the WINNMR format prior to using 1D WIN-NMR or 2D WIN-NMR.

Depending on the software used for data transfer and on the original NMR data format different procedures are possible:

Transfer using NFS (Network File Server):

- The most convenient way is to use the NFS protocol, which has the effect of the hard disk of the central file server appearing as a normal PC disk (NFS mounted disk). NFS requires an Ethernet adapter and the NFS software, not included with the WIN-NMR software, to be installed on your PC. File transfer and copy operations of files stored on a NFS mounted disk are accomplished in exactly the same manner, using the same WINDOWS file manager or WIN-NMR commands, as for files stored on the hard disk(s) of your PC.

  Note: Access time for files stored on a NFS mounted disk will usually be longer than for the PC's dedicated hard disk(s). The actual time will depend on the type of network and the amount of network traffic and consequently may vary during the course of the working day.

Transfer using FTP (File Transfer Protocol)

- As an alternative to NFS the Bruker software module GETFILE was developed to manage and simplify communication with computers. For data transfer between workstations, acting as file-servers, or Bruker spectrometer computers running under UNIX (INDY, ASPECT Station 1, ASPECT X32) and PCs, the GETFILE program provides a graphical user interface for transferring data via Ethernet using FTP software and via the serial port, using Kermit software. For direct data transfer between ASPECT 2000/3000 computers and PCs via the parallel port NMRLINK or FASTRAN software is used.

GETFILE is started with the **File Transfer** command in both 1D WIN-NMR and 2D WIN-NMR. For a transfer using FTP – the most common transfer protocol – the local and remote FTP must first be set up using the items in the **FTP** pull-down menu. After defining the file to be copied, a number of additional parameters must be set such as the type of data, the format (UXNMR/XWINNMR or WIN-NMR) to be used for the stored data and the destination directory. When all the appropriate parameters have been set, the transfer is initialised with the **Get NMR-Files ...** command.

For a more detailed description of this program the reader is referred to the GETFILE manual [2.3]. This manual contains useful information about building-up a local NMR data network, choosing the most appropriate transfer protocol and how to interface personal requirements with an existing local network.

It is the NFS based transfer and conversion of NMR data stored on remote computers which will be demonstrated in the following *Check its*.

2.7.2.1 UXNMR/XWINNMR-Format

For any UXNMR/XWINNMR format 1D or 2D NMR file stored on a remote computer the transfer via NFS, its conversion into the WINNMR format and its storage on the hard disk of your PC may be performed in either an automatic or interactive stepwise manner

For automatic data transfer and data conversion choose the **Open** option from the **File** pull-down menu in either 1D WIN-NMR or 2D WIN-NMR in exactly the same way as for data files already stored on your PC hard disk. **Open** directly transfers and automatically converts (UXNMR/XWINNMR-data) 1D or 2D data sets and shows the NMR data in the corresponding WIN-NMR application window for immediate processing. The processed data will not automatically be stored in the WINNMR format.

A first alternative for 1D WIN-NMR only is to use the **Filecopy & Convert** command in the **File** pull-down menu. This will transfer and store the selected data file in a predefined destination directory either in the WIN-NMR or the original UXNMR/XWINNMR format. If the **As Source** option is selected in the dialog box, the original UXNMR/XWINNMR file structure is created on the destination directory. This feature can be used to copy one or more UXNMR/XWINNMR directory structures to a PC disk from files stored either on the same disk or on NFS mounted disks. For further processing, files transferred using this method must be loaded from within 1D WIN-NMR using the **Open** command.

Attention: The **Filecopy & Convert** option is only available in 1D WIN-NMR and should not be used to copy or convert 2D files (see section 2.7.3).

A second alternative, available for both 1D WIN-NMR and 2D WIN-NMR, is similar to the **Filecopy & Convert** method just discussed. From the **File** pull-down menu choose the **File Transfer** option to load the GETFILE program. From the **File** pull-down menu in the GETFILE dialog box choose the **Copy & Convert UNIX-Files ...** option to transfer and automatically convert the selected file into the specified destination directory.

In the subsequent *Check its* the transfer and conversion of 1D and 2D NMR data files in the UXNMR/XWINNMR format is demonstrated using various alternatives. The converted data files will be stored and will be at your disposal for later processing in chapter 5.

Check it in 1D WIN-NMR:

If you have installed the NMR data on your hard disk, assume that your CD-ROM disk (denoted E:) is now a NFS mounted remote disk. Start the 1D WIN-NMR program and from the **File** pull-down menu choose the **Open** option to read in a 1D NMR data file in the UXNMR/XWINNMR format. The **Open** dialog box appears on the screen. Select the directory E:\NMRDATA\FORMAT\XWINNMR\1D\H\HUX on your CD-ROM disk, and mark the

UXNMR/XWINNMR file UX HUX 001 001 FID by clicking on it. Start the data conversion process by clicking on the **OK** button in the dialog box. An info-box appears on the screen indicating the current operation ("copy .... convert") on the data set being read and the FID is displayed on the screen.

To save the converted file, from the **File** pull-down menu choose the **Save as ...** option. Use the default name in the dialog box that appears on the screen C:\DEMO1D\DAT\ASPX32\HUX\001001.FID. Click on the **OK** button in this dialog box to initialise the save operation. The converted file will now be stored on your PC. Check the new entry in the corresponding directory.

Check it in 1D WIN-NMR:

Again assume that your CD-ROM disk E: is a NFS mounted remote disk. Start 1D WIN-NMR, from the **File** pull-down menu choose the **Filecopy & Convert** option. In the corresponding dialog box select the directory E:\NMRDATA\FORMAT\XWINNMR\1D\C\CUX with both the spectrum (1R) and the FID (FID) of the file CUX. Press the **CTRL**-key and select both files using the left mouse button and hit the **OK** button. Confirm in the second **Filecopy & Convert** dialog box the default destination directory C:\DEMO1D\DAT\ASPX32, define the destination format (**1D WINNMR**) and click on the **OK** button to initialize the copy and convert process. Check the presence of the corresponding two new WIN-NMR data files in this directory.

Check it in GETFILE:

An alternative way to copy and convert a 1D NMR data file in the UXNMR/XWINNMR format is to use the GETFILE module. Click either on the GETFILE icon in your WINDOWS program manager, or start GETFILE from within 1D WIN-NMR using the **File** pull-down menu to activate the **File Transfer** option. The GETFILE dialog box appears on the screen. From the **File** pull-down menu in this new dialog box, choose the **Copy & Convert UXNMR-Files ...** option. The **Convert Files** dialog box appears on the screen. Select the CD-ROM directory E:\NMRDATA\FORMAT\XWINNMR\1D\H\HUX, mark the UXNMR/XWINNMR file UX HUX 001 001 1R by clicking on it and define the destination directory (**Dest. Dir...**) as C:\DEMO1D\DAT\ASPX32. Start the data conversion process by clicking on the **OK** button in the same and the next dialog box. An info-box appears on the screen indicating the current operation ("copying UXNMR/XWINNMR files .... converting") on the data set being read.

Note: With this GETFILE variant whole series of data files in the UXNMR/XWINNMR format may be converted automatically. Press the **CTRL**-key on your keyboard while you select in the usual way the files you want to copy and convert with your mouse.

Exit the GETFILE module, go back to the 1D WIN-NMR program and check the new entry in the corresponding directory. There should now be the FID and the spectrum file W HUX 001 001 FID and W HUX 001 001 1R respectively.

Check it in 2D WIN-NMR:

Again assume that your CD-ROM disk is a NFS mounted remote disk. Start the 2D WIN-NMR program and from the **File** pull-down menu choose the **Open** option to read in a 2D NMR data file in the UXNMR/XWINNMR format. The **Open File** dialog box appears on the screen. Select the CD-ROM directory E:\NMRDATA\FORMAT\XWINNMR\2D\HH\HHUX. Mark the UXNMR /XWINNMR file UX HHUX 001 001 SER (the 2D raw data of a $^{1}H/^{1}H$-COSY experiment) by clicking on it. Start the data conversion process by clicking on the **OK** button in the dialog box. An info-box appears on the screen indicating the current operation ("copy .... temporary files, ... working, ...converting data ...working") on the data set being read. Horizontal bars in the lower part of this box monitors the progress of the various operations. Finally the message "2D Time Domain Data available Type ´xfb´ to process" appears.

To save the converted file, from the **File** pull-down menu choose the **Save as ...** option. In the dialog box that appears on the screen choose the name D:\NMRDATA\FORMAT\XWINNMR\2D\HH\HHUX\001001.SER and select the WINNMR format. Click on the **OK** button in this dialog box to initialize the save operation.

Check the new entry in the corresponding directory. There should now be two SER files, W HHUX 001 001 SER and UX HHUX 1 1 SER.

Check it in GETFILE:

An alternative way to copy and convert a 2D NMR data file in the UXNMR/XWINNMR format is to use the GETFILE. Click either on the GETFILE icon in your WINDOWS program manager, or start GETFILE from within 2D WIN-NMR using the **File** pull-down menu to activate the **File Transfer** option. The GETFILE dialog box appears on the screen. From the **File** pull-down menu in this new dialog box, choose the **Copy & Convert UXNMR-Files ...** option. The **Convert Files** dialog box appears on the screen. Select the CD-ROM directory E:\NMRDATA\FORMAT\XWINNMR\2D\HH\HHUX, mark the UXNMR/XWINNMR file UX HHUX 1 1 2RR by clicking on it and define the destination directory as D:\NMRDATA\FORMAT\XWINNMR\2D\HH. Start the data conversion process by clicking on the **OK** button in the same an the next dialog box. An info-box appears on the screen indicating the current operation ("copying UXNMR/XWINNMR files .... converting") on the data set being read.

Exit the GETFILE module, go back to the 2D WIN-NMR program, save the data and check the new entry in the corresponding directory. There should appear two spectra files, W HHUX 001 001 RR and UX HHUX 1 1 2RR.

For more details concerning the conversion of NMR data in the UXNMR/XWINNMR format the reader is referred to the information in the GETFILE [2.3], 1D WIN-NMR [2.1] or 2D WIN-NMR [2.2] manuals or in the on-line HELP routines of 1D- and 2D WIN-NMR.

2.7.2.2 DISNMR-Format

For any 1D or 2D DISNMR format file stored on a remote computer the transfer via NFS, its conversion to WINNMR format and its storage on the hard disk of your PC may be achieved in a number of ways. Files may be converted automatically with the minimum of user participation or in a stepwise interactive manner. The transfer and conversion procedures are different for 1D and 2D data files.

For 1D data files three possibilities using the **Open**, **Filecopy & Convert** or **File Transfer** 1D WIN-NMR commands are available and the procedures are the same as described above for UNIX format NMR data files.

For 2D data files in the DISNMR format file transfer via NFS must be accomplished with the GETFILE program, which may be loaded directly or using the **File Transfer** command in the 2D WIN-NMR **File** pull-down menu. Using the **Copy NMR Files ...** command in the corresponding **File** pull-down menu of GETFILE initialises the automatic transfer into the specified destination directory as a DISNMR format data file. To convert the 2D data file into the WINNMR format activate the **Convert ASPECT Files ...** command in the **File** pull-down menu. A dialog box will appear on the screen and a number of options may be selected within this box. Clicking on the **OK** button opens the **ST2D** dialog box containing various parameters that may be modified before the data is converted. Clicking on the **SER Format** button starts the data conversion. After being converted the data may be loaded from within the 2D WIN-NMR program for further processing.

In the subsequent *Check its* the transfer and conversion of 1D and 2D NMR data files in the DISNMR format is demonstrated. The converted data files will be stored and will be at your disposal for later processing in chapter 5.

Check it in 1D WIN-NMR:

Again assume that your CD-ROM disk E: is a NFS mounted remote disk. Start the 1D WIN-NMR program and from the **File** pull-down menu choose the **Open** option to read in a 1D NMR data file in the DISNMR format. The **Open File** dialog box appears on the screen. Select the CD-ROM directory E:\NMRDATA\FORMAT\DISNMR\1D\H, mark the DISNMR file DIS HDIS.001 by clicking on it and start the data conversion process by clicking the **OK** button in the dialog box. An info-box appears on the screen indicating the current operation on the data set being read and an FID is shown on the screen.

The NMR data will be converted and will automatically be saved in the WINNMR format as C:\DEMO1D\DAT\ASP3000\HDIS\001001.FID, but may also be stored under another name using the **Save as ...** option in the **File** pull-down menu. Make sure that the binary format has been selected in the dialog box appearing on the screen and click the **OK** button which will initialize the save operation.

Check the new entry in the corresponding directory. This converted data will be processed in chapter 5.3.2.

Check it in GETFILE:

As an alternative way to copy and convert a 1D NMR data file in the DISNMR format use the GETFILE module. Click either on the GETFILE icon in your WINDOWS program manager, or start GETFILE from within 1D WIN-NMR using the **File Transfer** option in the **File** pull-down menu. The **GETFILE** dialog box appears on the screen. From the **File** pull-down menu in this new dialog box, choose the **Copy NMR-Files...** option. Select the CD-ROM directory E:\NMRDATA\FORMAT\DISNMR\1D\C, mark the DISNMR file CDIS.001 and specify the destination directory, C:\DEMO1D\DAT\ASP3000, in the additional dialog box opened by hitting the **Dest. Dir...** button. Start the file transfer with the **OK** button in this and the next dialog box. Check the new entry (CDIS.001) in the corresponding directory.

To convert this DISNMR file activate again the GETFILE **File** pull-down menu and choose the **Convert ASPECT-Files ...** option. The **Convert ASPECT-Files** dialog box appears on the screen. Select the CDIS.001 file in the directory C:\DEMO1D\DAT\ASP3000, hit the **Select** button and start the data conversion process by clicking the **OK** button in the dialog box. An info-box appears on the screen indicating the current operation on the data set being read. Exit this dialog box clicking on the **Cancel** button. The converted file will be added as a directory C:\DEMO1D\DAT\ASP3000\CDIS.

Exit the GETFILE module, go back to the 1D WIN-NMR program and check the new entry in the WINNMR format in the corresponding directory. There should now be a FID file W CDIS 001001.FID. This converted data will be processed in chapter 5.3.2.

The 2D WIN-NMR program does not allow the direct transfer and conversion of a 2D data file in the DISNMR format. The GETFILE module is the only way to do the conversion.

Check it in GETFILE and 2D WIN-NMR:

Assume that your CD-ROM is a NFS mounted remote disk E:. Click either on the GETFILE icon in your WINDOWS program manager, or start GETFILE from within 2D WIN-NMR using the **File** pull-down menu to activate the **File Transfer** option. The **GETFILE** dialog box appears on the screen. From the **File** pull-down menu in this new dialog box, choose the **Copy NMR-Files...** option. Select the CD-ROM directory E:\NMRDATA\FORMAT\DISNMR\2D\HH, mark the DISNMR file HHDIS.SER (2D $^1H/^1H$-COSY experiment) and specify the destination directory, D:\NMRDATA\FORMAT\DISNMR\2D\HH, in the additional dialog box opened by hitting the **Dest. Dir...** button. Start the file transfer with the **OK** button in this and the next dialog box. Since the file HHDIS.SER had already been copied to your disk when installing the NMR data base, cancel the transfer and overwriting by hitting the **Cancel** button in the dialog box appearing on the screen.

To convert this 2D DISNMR file activate again the GETFILE **File** pull-down menu and choose the **Convert ASPECT-Files ...** option. The **Convert ASPECT-Files** dialog box appears on the screen. Select the HHDIS.SER file in the directory D:\NMRDATA\FORMAT\DISNMR\2D\HH, hit the **Select** and the **OK** button which opens the dialog box **ST2D:´HHDIS.SER´**. Check the parameters - the meaning of the most important ones will be explained in the course of this book - and change their values if necessary. Some of them – **ND0** (number of t1 periods in the corresponding 2D experiment), **SF (F1)** (spectrometer frequency valid for the second dimension F1 of the corresponding 2D experiment) – must be set correctly at this stage, others may be modified after the conversion process within 2D WIN-NMR. Neither **ND0** nor **SF (F1)** must be changed for this $^1H/^1H$-COSY data file. Click on the **SER Format** button which will start the conversion process and exit the **Convert Aspect Files** dialog box by clicking on the **Cancel** button.

The corresponding 2D data file in the WINNMR format has been stored as D:\NMRDATA\FORMAT\DISNMR\2D\HH\HHDIS\001001.SER. Check this with 2D WIN-NMR or with your WINDOWS file manager (WINDOWS Explorer). To start this file manager from within 2D WIN-NMR select **File Manager** in the **File** pull-down menu. The original DISNMR format 2D file, stored in the D:\NMRDATA\FORMAT\DISNMR\2D\HH directory is not visible from within 2D WIN-NMR.

The converted 2D data will be processed in chapter 5.3.2.

Check it in GETFILE and 2D WIN-NMR:

Repeat what you have just done, but select now the CD-ROM directory E:\NMRDATA\FORMAT\DISNMR\2D\CH\CHDIS. Again hit the **Cancel** button when asked for overwriting. Convert the DISNMR format 2D data file CHDIS.SER (2D $^1H/^{13}C$-COSY experiment) following the same guidelines as outlined above. Carefully check the parameters and change **NDO** from 1 to 2 and **SF(F1)** from 75.47 MHz to 300.13 MHz before the conversion is initialized by hitting the **Ser Format** button. It will not be possible to alter these two parameters after conversion to the WINNMR format.

The corresponding 2D data file in the WINNMR format has been stored as D:\NMRDATA\FORMAT\DISNMR\2D\CH\CHDIS\001001.SER. Check this with 2D WIN-NMR or with your WINDOWS file manager (WINDOWS Explorer). The original DISNMR format 2D file, stored in the D:\NMRDATA\FORMAT\DISNMR\2D\CH directory is not visible from within 2D WIN-NMR.

The converted 2D data will be processed in chapter 5.3.2.

For more details concerning the conversion of NMR data in the DISNMR format the reader is referred to the information in the GETFILE [2.3], 1D WIN-NMR [2.1] or 2D WIN-NMR [2.2] manuals or in the on-line **Help** routines of 1D- and 2D WIN-NMR.

2.7.3 Decomposition of 2D Data Files

The 1D WIN-NMR **Filecopy & Convert** option, besides being used for 1D data files as described above, has a further important processing function for 2D NMR data. When **Filecopy & Convert** is applied to a 2D file, either the raw data (SER) or the processed Fourier transformed spectra (2RR/2II), it copies and/or decomposes (splits) it into a series of 1D FIDs or spectra respectively. The original data may be in WINNMR, UXNMR/XWINNMR or DISNMR format. This option was developed to have easy and most convenient access to NMR data originating from "pseudo-2D" experiments, i.e. serial measurements such as the 1D T_1 Inversion Recovery or the 1D NOE experiment, discussed in chapter 3 and performed in a 2D mode on the spectrometer. This function is especially helpful if you have direct access to the spectrometer's data system or a central server, where the original data is stored. The 2D files may be decomposed into several 1D files with ascending experiment numbers. The path for saving this data on the PC is derived from the file name and the "work" entry in the [nmr-directories] section of the DEMO1D.INI file *e.g.* \<WORK>\<NAME>. If WIN-NMR has been specified as the destination format, the series of 1D FIDs generated using **Filecopy & Convert** are easily processed and converted into spectra using the 1D WIN-NMR **Serial Processing** function (see chapter 5). For UXNMR/XWINNMR data, the original 2D file will be replaced by the series of 1D files in the WINNMR format and will be no longer available.

In conclusion, **Filecopy & Convert** enables you in a very simple manner to access and process the data, either FIDs or spectra, from relaxation, NOE or other experiments performed in the "pseudo-2D" mode directly in 1D WIN-NMR.

Check it in 1D WIN-NMR:

Start 1D WIN-NMR, from the **File** pull-down menu and choose the **Filecopy & Convert** option. Select in the corresponding dialog box in the directory

D:\NMRDATA\GLUCOSE\1D\C\GCT1\2D the "pseudo 2D" file W 2D\001001. SER as obtained from a T_1 Inversion Recovery experiment. Hit the **OK** button and define in the second **Filecopy & Convert** dialog box the destination directory D:\NMRDATA\GLUCOSE\1D\C\GCT1 and the destination format (1D WINNMR) and click on the **OK** button to initialize the decomposition process. Check that the 2D file will be decomposed in a series of 1D files under the same name (e.g. 2D\001001.FID, 2D \002001.FID, ...) in the directory D:\NMRDATA\GLUCOSE\1D\C\GCT1\2D. This same series has already been copied into the directory D:\NMRDATA\GLUCOSE\1D\C\GCT1\1D (e.g. 1D\001001.FID, 1D\002001.FID, ...) when installing the NMR data base on disk D: of your PC.

2.8 References

[2.1] Bruker, *1D WIN-NMR Manual*, Release 960901

[2.2] Bruker, *2D WIN-NMR Manual*, Release 950901

[2.3] Bruker, *GETFILE Manual*, Release 950901

Please contact your next Bruker representative or

Bruker-Franzen Analytik GmbH
Fahrenheitstr. 4
D-28259 Bremen

FAX No: 0049-421-2205102

to order these manuals.

3 Modern Homo- and Heteronuclear 1D and 2D NMR Experiments: A Short Overview

3.1 Introduction

In section 3.2 the main principles of 1D and 2D NMR experiments are briefly discussed and some typical examples are shown. It is not the aim of this book to give you an introduction into the mechanics of multiple pulse experiments or the "gymnastics of spins" in such experiments. If you are interested in the physics and the experimental aspects of the various NMR experiments you are referred to *Data Acquisition* (volume 2) of this series, where these topics are discussed in detail.

Section 3.3 gives an overview of some of the most useful 1D- and 2D NMR experiments used in unravelling molecular structures. Each experiment is shown in its most simple and basic version, although more sophisticated schemes have been used to acquire the corresponding spectra. The discussion for each experiment is divided into four parts:

- Theory – A short description of the experiment and its theoretical background
- Pulse Diagram – A graphical representation of the pulse scheme
- Application – How the experiment may be used to solve structural problems, its advantages and limitations
- Example – A typical spectrum

NMR experiments may be grouped according to different characteristics. In the following sections the more methodical attributes of an NMR experiment have been used, for ordering the experiments. The same attributes determine the hierarchy of the directory chosen for the NMR data of the peracetylated β-D-glucose and the peracetylated oligosaccharide stored on CD-ROM.

It is the aim of this presentation, to demonstrate the most popular experiments used for measuring various nuclear properties, e.g. different types of spin-spin interactions (scalar/dipolar coupling). The chosen experiments use a variety of acquisition, detection and processing methods and include:

- Homonuclear and heteronuclear experiments
- Non-selective and selective excitation
- Selective perturbation/excitation using selective pulses or continuous irradiation

- 1D- and 2D experiments
- Normal and inverse mode of detection
- Magnitude and phase sensitive mode of detection
- Phase cycling and magnetic field gradients for coherence selection
- The use of double quantum or BIRD filters

If you are already familiar with the basic theory and with the application of modern NMR experiments you may want to skip this chapter and directly move to chapter 4.

3.2 The NMR Experiment

Any NMR experiment consists of a series of pulses and delays. Pulses are applied to perturb the thermal equilibrium of an ensemble of magnetically active spins and to force the spins to "speak" in a controlled and synchronised way. The evolution of these "spins conversations" , i.e. the evolution of coherences to be more precise, occurs in the intermediate delays and is manipulated by the pulses in the course of the pulse sequence. The spins response at the end of the pulse sequence is detected in a final detection or acquisition period. The general scheme for any pulse experiment is shown below:

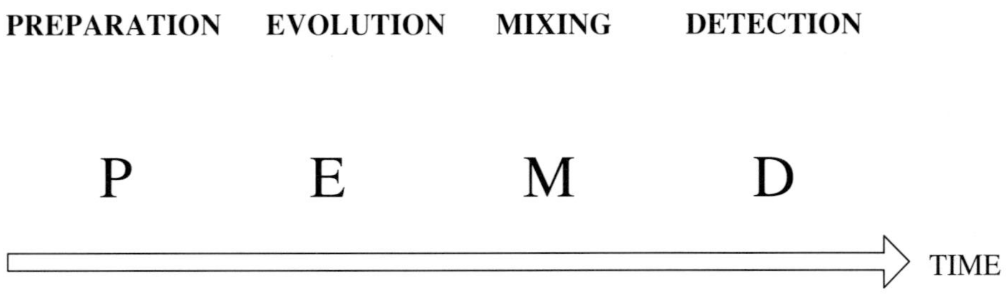

The pulse sequence starts with a preparation period **P**, which usually allows the ensemble of spins – still partially perturbed by the pulses applied in the preceding scan – to return back to the equilibrium state. The preparation period may also be used to force this ensemble of spins to a defined non-equilibrium state according to the operators needs.

After one or several pulses, the initial longitudinal magnetizations (polarizations) of an ensemble of spins are transferred either fully or partially into transverse magnetizations (coherences) which evolve during the evolution period **E** under the influence of several internal and external constraints. In the subsequent mixing period **M** the spins are allowed to "communicate" and to mutually exchange informations via several mechanisms such as polarization transfer, cross polarization or cross relaxation. The final detection period **D** is used to acquire the response of the spin system at the end of the pulse sequence. The very weak radio frequency signals or free induction decays

(FIDs) emitted by the nuclei are amplified over several stages, their frequencies are transformed from the MHz to the KHz range, unwanted signals are filtered out and the wanted signals are digitised using a suitable analogue to digital converter (ADC). The digitised raw data (FID) is finally stored on the spectrometers hard disk ready for data processing. The resonances in the corresponding spectra obtained after appropriate data processing are characterized by their frequencies, intensities, multiplicities and by their linewidths. These properties are influenced by a series of structure dependent parameters such as chemical shifts, the number of nuclei involved, scalar and dipolar coupling, by molecular mobilities, but also by external factors such as the homogeneity of the static magnetic field or the intensity and frequency of additional radio frequency sources.

In contrast to the preparation and the detection period which are part of any pulse sequence, the evolution and the mixing period are not necessary in all types of NMR experiments.

Three typical examples of NMR pulse experiments are shown in Fig. 3.1:

A: In the simple homonuclear one pulse experiment, for the measurement of 1D spectra, the response of the spin system following a strong non-selective radio frequency pulse is acquired. The pulse sequence consists of a preparation and detection period only.

B: This pulse sequence, the 1D DEPT (Distorsionless Enhancement by Polarization Transfer) experiment, was developed to measure carbon chemical shifts with enhanced sensitivity and to determine at the same time their multiplicities, to differentiate between CH_3, CH_2, CH and C_q. It is a heteronuclear multiple pulse experiment with pulses applied to perturb both carbon and proton spins. It consists of a preparation, a mixing (used to transfer proton polarization to the directly bound carbons) and a detection period.

C: This pulse sequence, the popular homonuclear 2D COSY (COrelation Spectro-scopY) experiment, was designed to determine the entire $^1H/^1H$-coupling network of a molecule within a single experiment. The sequence consist of all four elements, i.e. a preparation, a evolution, a mixing and a detection period. The evolution period serves to introduce the second time (t1) domain of a 2D experiment and in the mixing period, which is actually a pulse, polarizations are exchanged among the coupled spins.

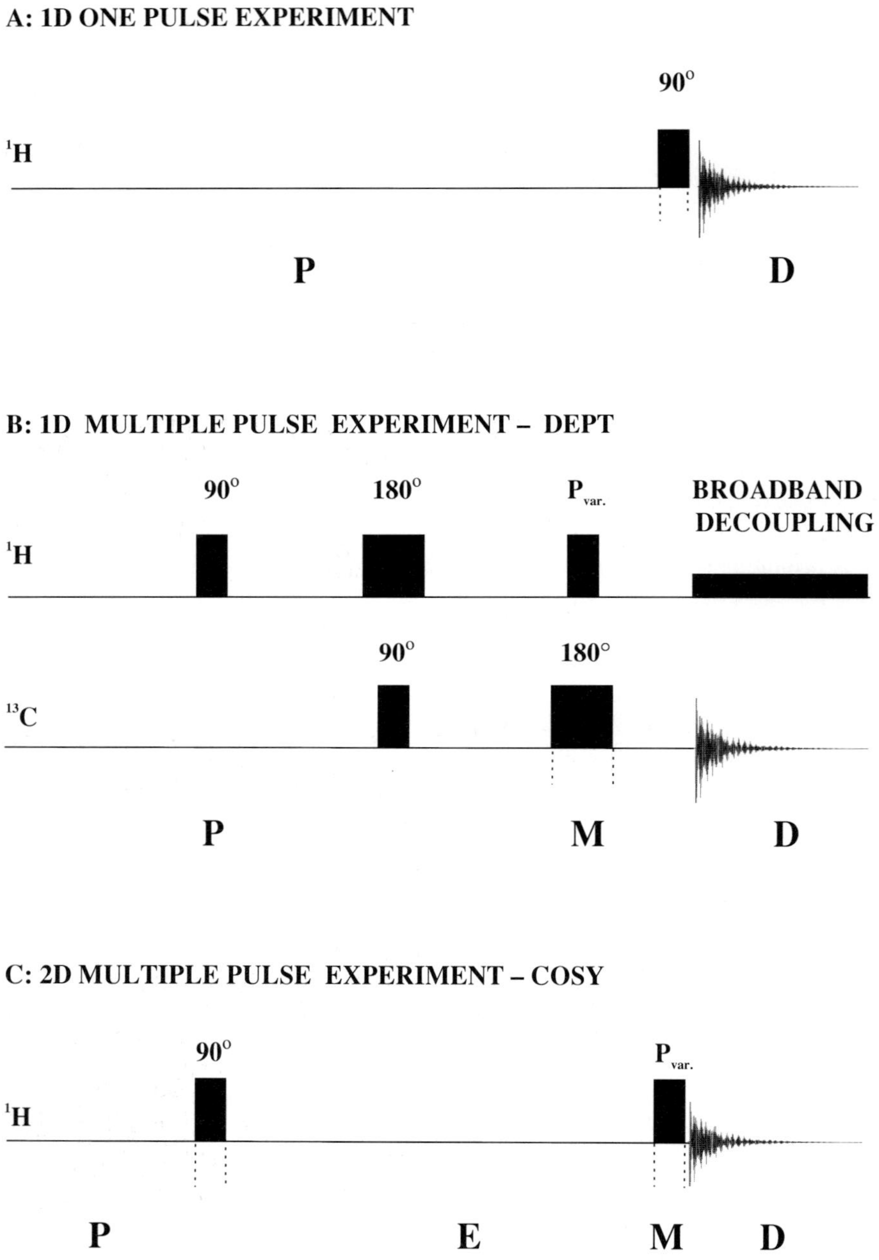

Fig. 3.1: Typical representatives of modern NMR experiments. Preparation, evolution, mixing and detection periods are abbreviated by **P**, **E**, **M** and **D** respectively. 180° and 90° pulses are indicated. $P_{var.}$ denotes a pulse of variable length.

3.3 1D Experiments

3.3.1 ^{1}H Experiments

3.3.1.1 ^{1}H One Pulse Experiment

Theory

The simplest and most often applied NMR experiment is the one pulse ^{1}H experiment shown in Fig. 3.2. The pulse sequence consists of the recycle delay D1 (preparation period) followed by a radio frequency (rf) pulse P1. The pulse excites all the proton spins of a molecule and generates transverse magnetization (coherences) evolving in time and carrying for each spin information concerning its chemical shift as well as its scalar couplings to other spins and its relaxation properties. Data is collected following the rf pulse (detection period). The length of this detection period is denoted as acquisition time. In practice a value for P1 close to 90° is normally used and the recycle delay D1 is set long enough to avoid problems with partial saturation.

Note that the time intervals depicted in this as well as in the other pulse sequence diagrams in this chapter, are not drawn to scale. Similarly, the pulse (hundreds of volts peak to peak) and the detected signal (microvolts) are also not drawn to scale.

Pulse Diagram

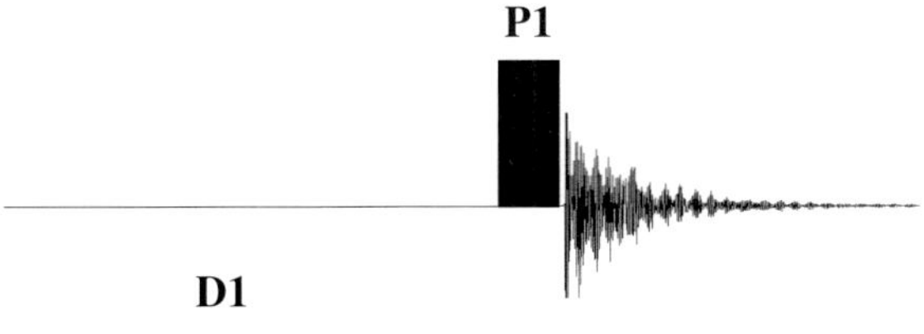

Fig. 3.2: The one pulse experiment

Application

The experiment is used for solving simple structural problems, to check the progress of synthetic work and for setting up subsequent, more sophisticated experiments. Chemical shifts and coupling constants may be evaluated or – for more complex spectra showing strong signal overlap and/or strong coupling effects – at least estimated. For the precise evaluation of these parameters additional tools for deconvolution (1D WIN-NMR and WIN-FIT) or for simulation/iteration (WIN-DAISY) are available.

The integration of the individual proton resonances yield the ratios of the numbers of corresponding protons. This information is helpful for signal assignments and, as in the case of mixtures, for quantitative analysis.

Example

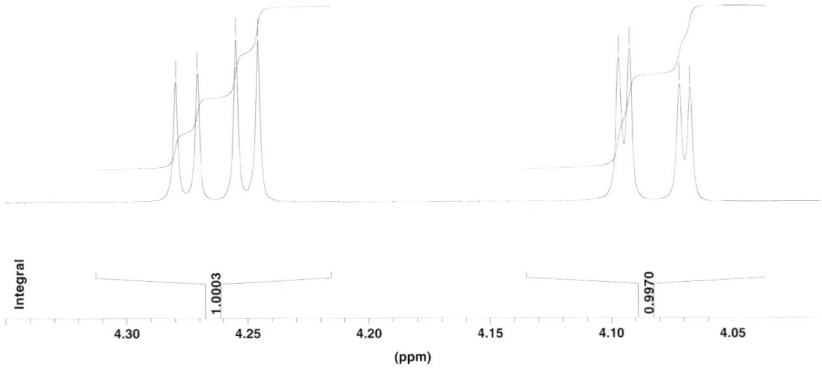

Fig. 3.3: Expansion of a ^{1}H spectrum with integral traces and integral values

3.3.1.2 ^{1}H {^{1}H} Selective Decoupling Experiment [3.1]

Theory

In this homonuclear double resonance experiment one proton or one group of protons is selected for selective decoupling to remove its scalar coupling effect from other protons within the molecule. Decoupling is usually applied throughout the whole experiment, including the detection period. The pulse sequence (Fig. 3.4) consists of the recycle delay D1 followed by an rf pulse P1 and a continuous weak rf-decoupling field (continuous wave decoupling) at a selected frequency applied on a second (decoupler) channel. To prevent a large spike appearing in the spectrum at the selected frequency, the decoupler and the receiver are alternatively gated on and off in the detection period.

Pulse Diagram

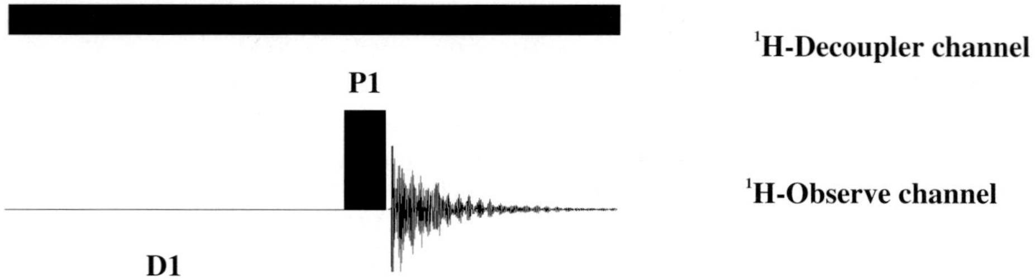

Fig. 3.4: One pulse experiment with selective homonuclear decoupling

Application

The experiment is used for solving simple structural problems and to prove small number of scalar coupling interactions. To identify a large number of scalar coupling interactions, the COSY experiment (section 3.4.1.1) may be a more efficient way but the exact method will depend upon the nature of the problem. Selective homonuclear decoupling is also used to simplify a complex multiplet pattern for subsequent spectral analysis by cancelling one of the coupling interactions to the observed resonance. When analysing such spectra you should be aware of the slight changes in the chemical shifts and reductions in coupling constants. These "Bloch-Siegert Shifts" are caused by the decoupling field and the magnitude of the changes are dependent upon its strength and its frequency relative to the involved resonances.

Example

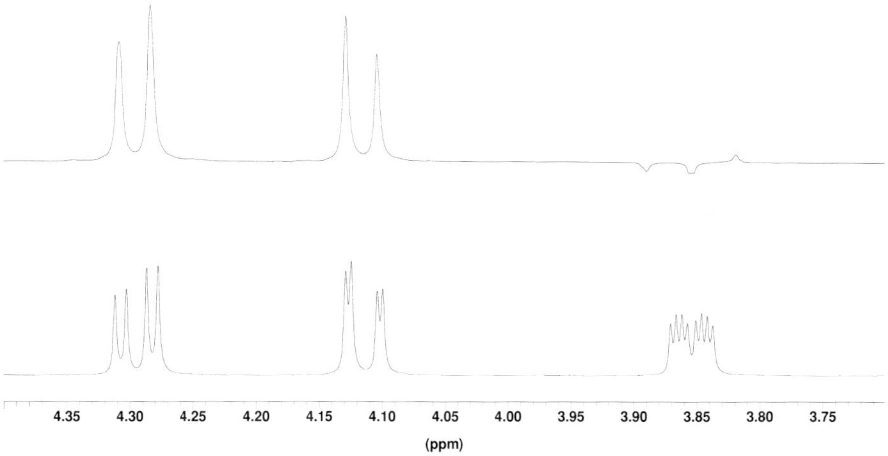

Fig. 3.5: Selective decoupled spectrum (top trace) and normal spectrum (bottom trace). The spin with resonance at 3.85 ppm has been decoupled.

3.3.1.3 $^1H\{^1H\}$ Total Correlation Spectroscopy (TOCSY) Experiment [3.2]

Theory

This ^1H experiment (Fig. 3.6) was designed to selectively excite an ensemble of coupled spins, J-isolated from other spin ensembles, and to measure the corresponding subspectrum ("spin chromatography"). This is accomplished by selectively exciting one spin of the ensemble with a selective 90° pulse at the beginning of the sequence; cross polarization then distributes this perturbation during a "spin-lock" period step wise among all the coupled spins of the selected spin system. This is in sharp contrast to the former selective decoupling experiment where the selective perturbation affects only and exclusively those spins which are directly coupled to the perturbed spin(s). The spin lock is generated by a series of strong rf pulses of different pulse lengths and follows a complex phase scheme (e.g. MLEV). To a first approximation, the extent of the propagation of cross polarization through the spin system depends upon the length of the mixing period.

Since a single experiment with a fixed spin-lock period is usually performed, this experiment does not, in contrast to the $^1H/^1H$ decoupling experiment, establish the complete coupling network. For this purpose a series of 1D TOCSY experiments, with the length of the mixing period increasing step-wise from experiment to experiment, has to be performed.

To investigate the subspectra of several isolated spin systems, it is necessary to perform a whole series of experiments using the appropriate target spins for selective perturbation.

The shape of the resonance signals in this experiment are no longer in pure absorption. To minimise these unwanted effects "z-filter"-TOCSY experiments are applied.

Pulse Diagram

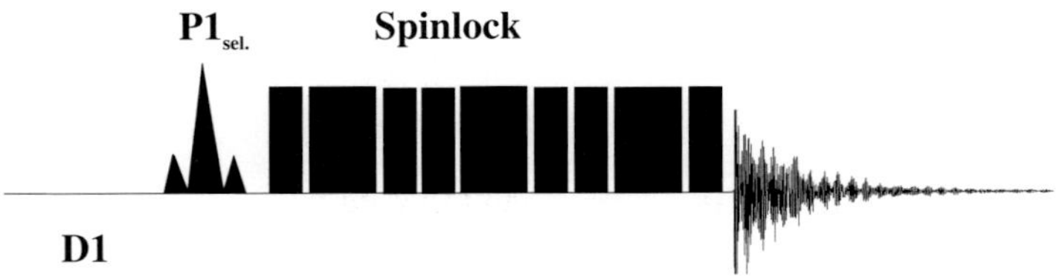

Fig. 3.6: The 1D TOCSY experiment

Application

The data of 1D TOCSY experiments yield subspectra of coupled spin systems. This is particularly helpful in cases where part of the corresponding signals are covered by signals of other subspectra and where 1D homodecoupling experiments are not viable. The 1D TOCSY experiment is best suited to the investigation of molecules that consist of rows or a network of similar fragments with no coupling interactions between them, *e.g.* oligosaccharides, oligopeptides or oligonucleotides. In these cases, it allows the subspectra of the individual "monomers" to be extracted and analysed separately. Another field of application are mixtures of compounds where this method offers the possibility to artificially obtain spectra of "pure" compounds.

TOCSY spectra usually give no information about direct coupling interactions and hence details of J-coupling networks. This missing information is commonly obtained either from 1D homodecoupling or more efficiently from 2D COSY spectra.

Example

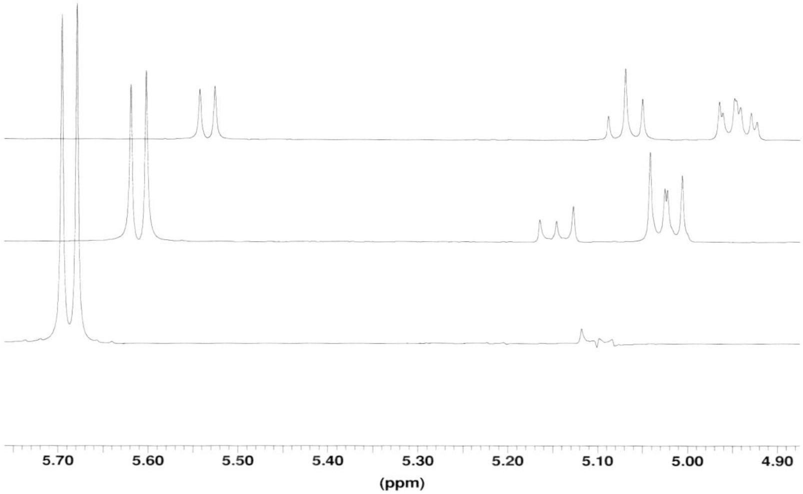

Fig. 3.7: 1D TOCSY spectra with long (top trace), medium (central trace) and short mixing times (bottom trace). The spin with resonance at 5.7 ppm has been selectively perturbed.

3.3.1.4 ^{1}H {^{1}H} Nuclear Overhauser (NOE) Experiment [3.3]

Theory

In this homonuclear double resonance experiment (Fig. 3.8) one proton or one group of protons is selected for selective perturbation prior to the acquisition time. During the irradiation period D1, a weak, selective and continuous rf field is applied. NOEs are built up due to dipolar spin-spin interactions (cross relaxation) for all those protons positioned closely in space to the perturbed proton(s). Depending on the delay D1, transient or the 0more intense "steady-state" (D1 greater than 5xT_{1max}) NOEs are obtained. The NOE depends on molecular properties such as internuclear distances and tumbling rates. According to theory NOEs are positive for small, highly mobile molecules (i.e. typical "organic" molecules) and are negative for large slowly reorienting molecules (i.e. biomolecules). As a consequence NOEs may be very small or even disappear in the worst case for molecules of intermediate size or for small molecules in rather viscous solutions, even if the corresponding protons are close in space. In such cases the 1D ROE experiment (section 3.3.1.5) is recommended as an alternative.

Besides the detection of NOEs this experiment also yields information about "exchanging" protons in a molecule. If an exchanging proton or a group of protons is preirradiated and saturated, part of this saturation is transferred with the proton from its original site in the molecule to the other sites(s) (saturation transfer). Since these exchange connected sites have in most cases different structural neighbourhoods, the corresponding resonances have usually different chemical shifts and the exchange process may easily be detected. The extent of saturation transfer depends on the type and rate of exchange and the relaxation rates of the corresponding protons.

NOE measurements are usually performed as a series of experiments with selected target spins to be irradiated; at least one experiment (used as reference in the subsequent data processing) is performed with the decoupler frequency set far removed from any proton resonance. The reference FID is subtracted from the FIDs obtained with selective perturbation of a proton or a group of protons prior to further processing. Fourier transformation results in a series of so-called "difference spectra" where even very small NOEs are easily identified.

The 1D NOE experiment is not restricted to the homonuclear $^1H/^1H$-case, but may also be performed in a heteronuclear mode, i.e. by observing NOEs for ^{13}C nuclei induced by selectively preirradiated protons.

Pulse Diagram

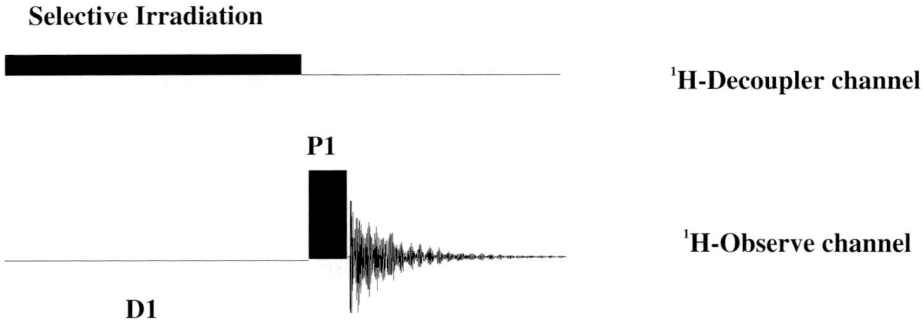

Fig. 3.8: The 1D ("steady-state")-NOE experiment

Application

The data of 1D NOE experiments yield qualitative or quantitative information on proton-proton distances within or between molecules. The experiment is used for detecting one or a few dipolar proton-proton interactions and is especially useful in solving stereochemical problems. If a large number of dipolar interactions or even all protons in a molecule are of interest, it is more efficient to use the 2D NOESY or 2D ROESY experiment (section 3.4.1.3). The 2D method is hampered however by the fact that only the inherently less intense transient rather than the stronger "steady-state" effects can be measured.

It is the combination of complementary structural information from NOE measurements together with the J-coupling information obtained from other experiments which makes NMR so successful.

Furthermore this experiment may also be used to detect and analyse exchange processes either on a qualitatively or a quantitatively basis.

Example

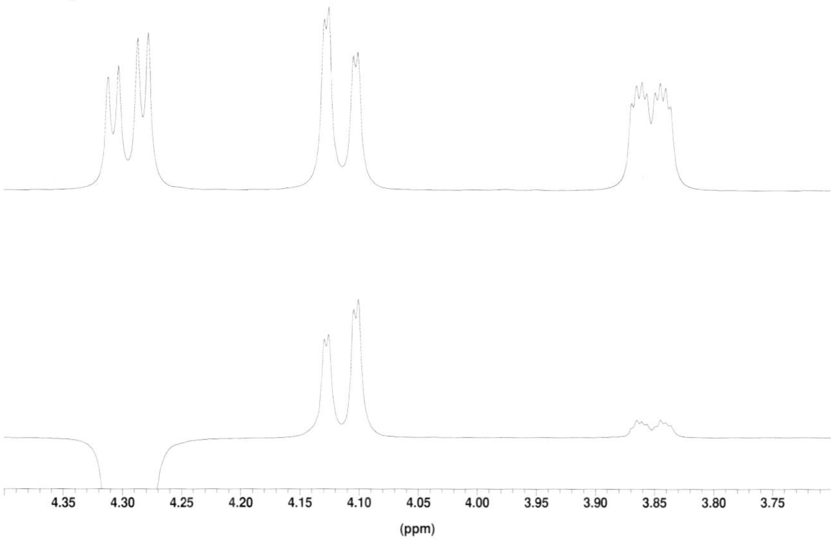

Fig. 3.9: 1D NOE Spectra: Top trace – Unperturbed reference spectrum. Bottom trace – Difference spectrum. The proton with resonance at 4.3 ppm has been selectively saturated.

3.3.1.5 ^{1}H {^{1}H} Nuclear Overhauser Experiment in the Rotating Frame (ROE) [3.4]

Theory

This ^{1}H experiment (Fig. 3.10) serves a similar purpose as the NOE experiment. In the NOE experiment, relaxation processes occur in the presence of the strong static magnetic field and a weak selective rf field. The ROE experiment is based upon cross relaxation processes ($T_{1\rho}$) observed between spins, that occur in the presence of a transverse, weak "spin-locking" rf field (either a continuous CW rf field or a series of weak rf pulses). According to theory, and in contrast to NOEs, ROEs are positive irrespective of the size or mobility of molecules and no unwanted "zero-passing" of the effect exists. However the effects at the small and large molecular size limits are both smaller compared to the corresponding NOE values of 0.5 and –1 respectively. The ROE experiment is ideal for intermediate sized molecules where NOEs may be close or equal to zero.

As with the NOE experiment, one proton or a group of protons is selected for selective perturbation prior to the acquisition time. Among several variations the simplest is the one using an initial selective 90° . A series of experiments with selected target spins to be irradiated is performed; in at least one experiment (used as reference in the subsequent data processing) the frequency of the selective 90° pulse is set well away from any proton resonance. Again, the reference FID is subtracted from the FIDs obtained with selective perturbation of a proton or a group of protons prior to further processing. Fourier transformation results in a series of difference spectra where even very small ROEs are easily identified.

Pulse Diagram

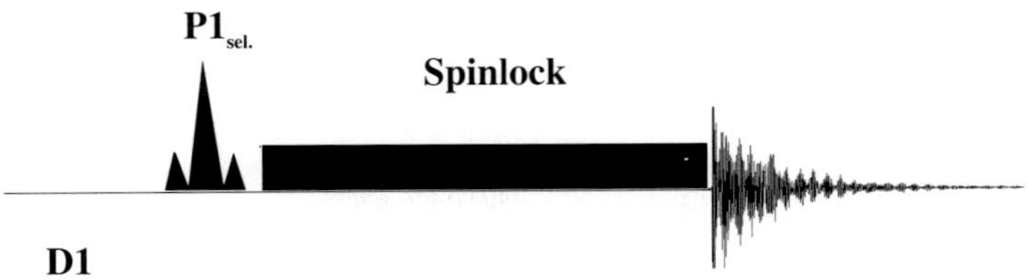

Fig. 3.10: The 1D ROE experiment

Application

The data, application and limitations of a 1D ROE experiment are similar to that obtained from a 1D NOE experiment (section 3.3.1.4). The 2D analogue ROESY is discussed in section 3.4.1.3.

Example

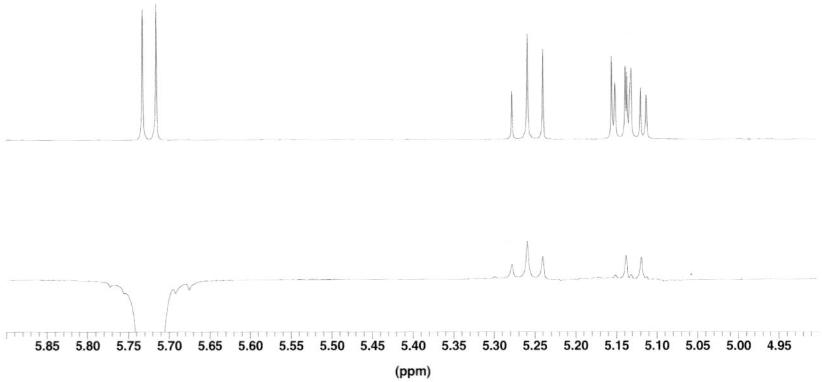

Fig. 3.11: 1D ROE Spectra: Top trace – Unperturbed reference spectrum. Bottom trace – Difference spectrum. The spin with resonance at 5.7 ppm has been selectively perturbed.

3.3.2 ^{13}C Experiments

3.3.2.1 ^{13}C One-Pulse Experiment

Theory

The one-pulse sequence is identical to the basic ^{1}H experiment, except that the rf pulse is applied at the ^{13}C frequency and that throughout the duration of the pulse sequence broadband ^{1}H decoupling is used to remove all heteronuclear J-coupling. The pulse sequence (Fig. 3.12) consists of the recycle delay D1 followed by an rf pulse P1. The pulse excites all ^{13}C spins of a molecule and generates transverse magnetizations (coherences) evolving in time and carrying for each spin chemical shift information. Data is collected following the rf pulse.

Since longitudinal relaxation and heteronuclear NOEs affect ^{13}C signal intensities and since the corresponding T_1 relaxation times and heteronuclear NOEs values vary for the different types of ^{13}C nuclei in a molecule a compromise must be found for the experimental parameters D1 and P1. To enhance the signals of the slowly relaxing quaternary carbon spins, large values of D1 and pulse angles less than 90° are chosen. The choice of D1 and P1 will not alleviate however, the difference in NOE values. For this reason quantitation by integration is usually not applied, but is possible if a modified experiment is performed and a few boundary conditions are meet.

Pulse Diagram

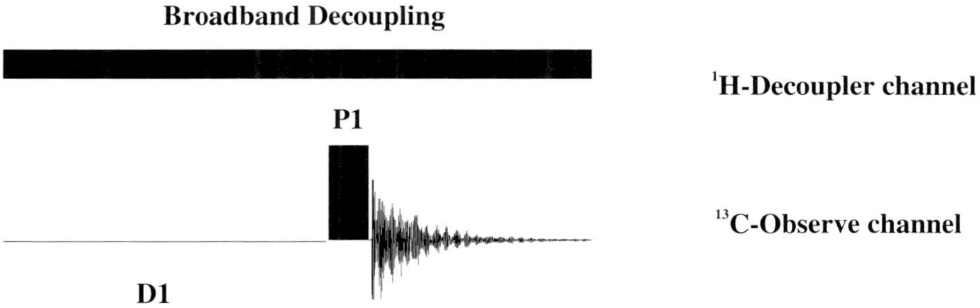

Fig. 3.12: The one-pulse ^{13}C experiment

Application

The experiment is used for solving simple structural problems and for the evaluation of chemical shifts. This experiment is usually combined with the DEPT experiment (see 3.3.2.2) for additional information and for signal assignments.

Example

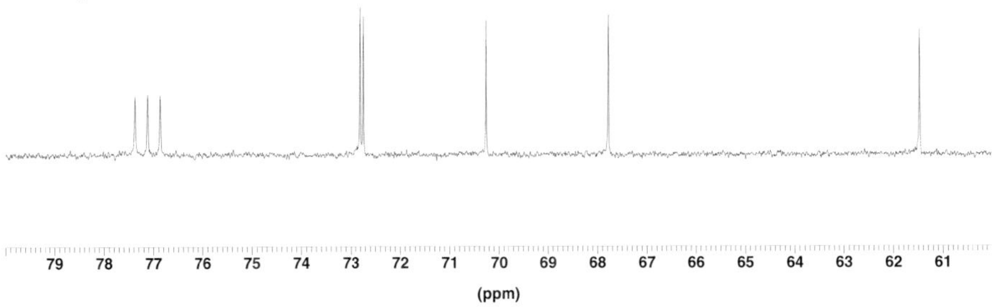

Fig. 3.13: A typical one-pulse ^{13}C spectrum

3.3.2.2 ^{13}C DEPT Experiment [3.5]

Theory

Distorsionless Enhancement by Polarization Transfer (DEPT) is a polarization transfer technique, exploiting the higher ^{1}H polarization and usually shorter ^{1}H T_1 relaxation times, and is useful for the observation of low-γ nuclei (commonly ^{13}C) which are J-coupled to ^{1}H. DEPT is also a spectral editing sequence, and may be used to generate separate ^{13}C subspectra for methyl (CH_3), methylene (CH_2), and methine (CH) signals. The delay D2 (see Fig. 3.14) between pulses on both the ^{13}C and the ^{1}H channel is adjusted to $1/(2^{1}J_{CH})$. The pulse angle (θ) of the final ^{1}H pulse P0 is the basis of spectral editing; with θ=45° the signals of all carbon multiplicities are visible with positive intensity, with θ=90° only the signals of methylene carbons are visible and with θ=135° again the signals of all carbon multipliciteis are visible, with positive intensities for CH and CH_3 groups and with negative intensities for CH_2 groups. Quaternary carbons are not observed in a DEPT spectrum.

DEPT is usually performed with broadband ^{1}H decoupling. It is relatively insensitive to the precise matching of delays with coupling constants, and so is much easier to use than the closely related INEPT or the JMOD (APT) (see section 3.3.2.3) sequence. DEPT, on the other hand, is more sensitive to pulse imperfections than INEPT or JMOD.

Pulse Diagram

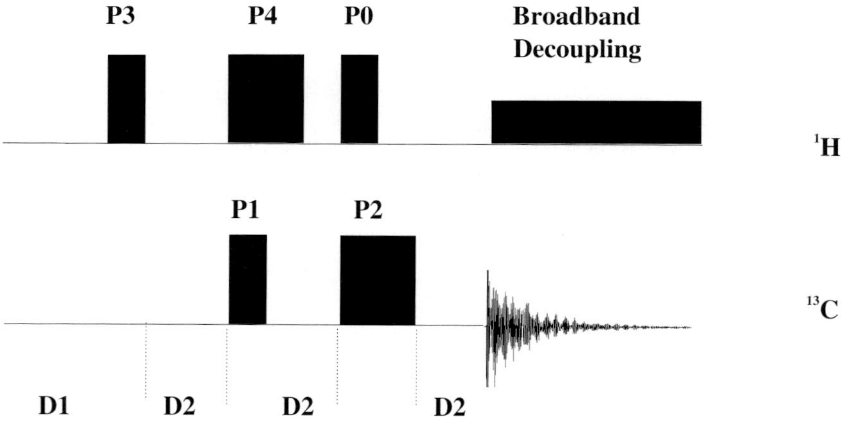

Fig. 3.14: The DEPT pulse sequence

Application

The experiment is used for solving simple structural problems, for the evaluation of chemical shifts and the determination of the multiplicities of the individual carbon signals. Special processing (see chapter 5) generates CH, CH_2 or CH_3 subspectra. This experiment is usually combined with the basic ^{13}C one pulse experiment to obtain the signals from quaternary carbons as well.

Example

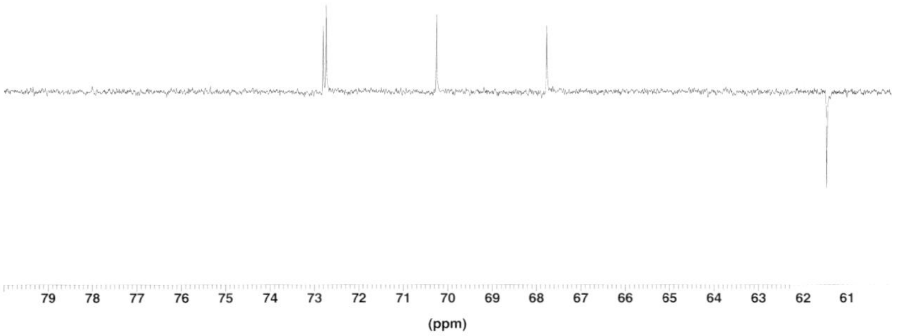

Fig. 3.15: DEPT spectrum with P0 set to 135° for θ

3.3.2.3 ^{13}C JMOD (APT) Experiment [3.6, 3.7]

Theory

The J-MODulated (JMOD) ^{13}C experiment, also known as Attached Proton Test (APT) was the first and simplest way to determine ^{13}C multiplicities. In contrast to DEPT no polarization transfer is included in the pulse sequence (Fig. 3.16) and as a consequence the signals of quaternary carbons are visible in the spectrum, but the sequence is far less sensitive than DEPT or INEPT. The value of D2 is used to differentiate between the different carbon multiplicities. The signal intensities of quaternary carbons are not influenced by the value of D2; for D2 equal to $1/^1J_{CH}$, CH and CH_3 groups have maximum negative intensity and CH_2 has maximum positive intensity. For D2 equal to $1/(2^1J_{CH})$ only the signals of quaternary carbons are visible.

JMOD (APT) is usually performed with broadband ^{1}H decoupling and is relatively sensitive to the precise matching of the delay D2 to the $^1J_{CH}$ coupling constant, and so is less easier to use than the polarization techniques DEPT and INEPT. On the other hand, only one single experiment is necessary to measure the signals of all carbon multiplicities.

Pulse Diagram

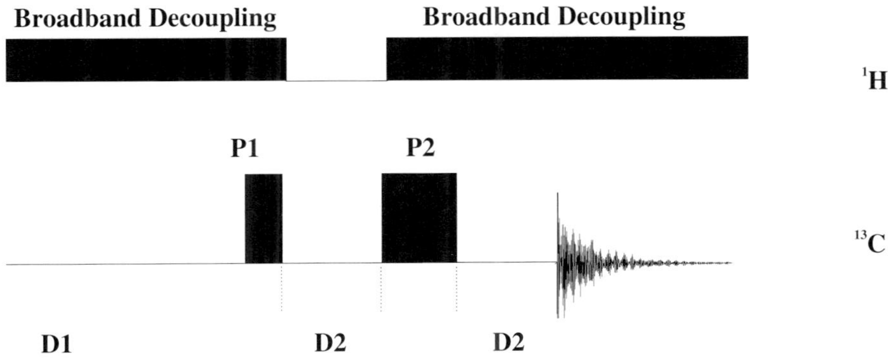

Fig. 3.16: The JMOD (APT) pulse sequence

Application

The experiment is used for solving simple structural problems, for the evaluation of chemical shifts and the determination of the multiplicities of individual carbon signals (including quaternary centres).

Example

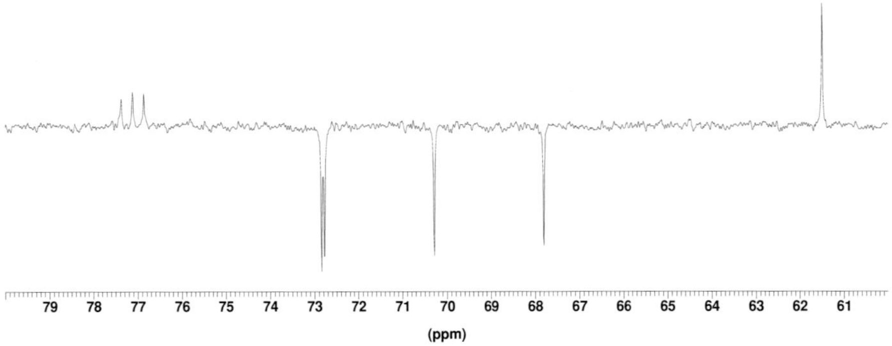

Fig. 3.17: The ^{13}C JMOD (APT) spectrum with D2 set to $1/^1J_{CH}$. Note that the $CDCl_3$ triplet is visible.

3.3.2.4 ^{13}C T_1 Inversion-Recovery Experiment [3.8, 3.9]

Theory

The ^{13}C T_1 Inversion-Recovery experiment is used to determine the longitudinal relaxation times, T_1. The pulse sequence (Fig. 3.18) starts, after a suitable preparation period D1 (D1 greater than $5xT_{1max.}$) to allow the spin system to reach thermal equilibrium, with a 180° pulse inverting all the carbon polarizations. The individual carbon spin ensembles return back to thermal equilibrium at different rates characterized by their T_1 values. This process is monitored through the delay D9, varied from experiment to experiment. The final 90° pulse generates transverse magnetizations. ^1H broadband decoupling is applied throughout. From the series of spectra obtained the T_1 values for each carbon may be evaluated by using the T_1 analysis module available with 1D WIN-NMR. The procedure is described in detail in *Modern Spectral Analysis* (volume 3 of this series).

The T_1 Inversion-Recovery experiment is not restricted to ^{13}C nuclei, but may also be applied to other nuclei, e.g. protons. In this case, the pulse sequence for the observe channel is the same, but no broadband decoupling is used.

Pulse Diagram

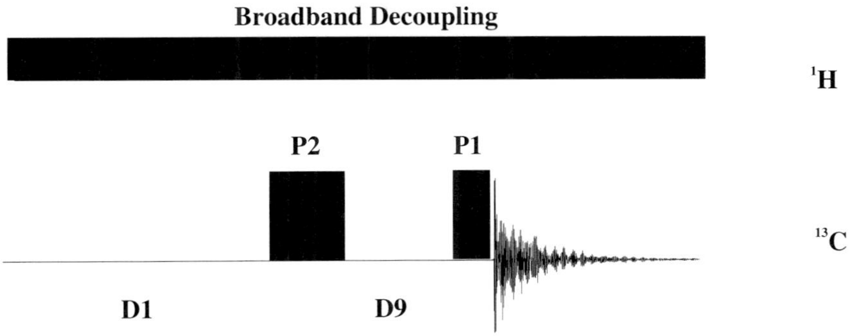

Fig. 3.18: The Inversion-Recovery pulse sequence for measuring T_1.

Application

The experiment is applied for the evaluation of ^{13}C T_1 values. T_1 values are usually used to optimize insensitive ^{13}C experiments, i.e. to adjust the length of the preparation time in other NMR experiments. To deduce structural information it is usual to interpret the dipolar part of the longitudinal relaxation time (T_1^{DD}). To separate the dipolar contribution from the contributions of other relaxation mechanisms, it is necessary to perform further experiments (gated decoupling experiments) to evaluate the heteronuclear NOE values. T_1^{DD} may be exploited in a qualitative way to differentiate between carbon nuclei in less or highly mobile molecular fragments. In a more detailed analysis reliable T_1^{DD} values can be used to describe the overall and internal motions of molecules.

Example

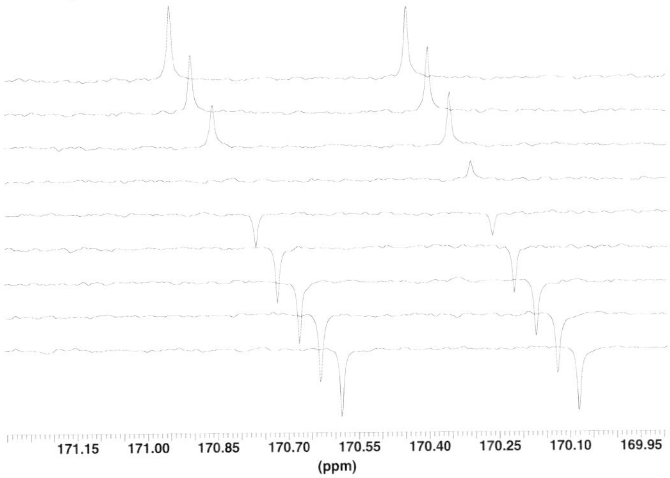

Fig. 3.19: Stacked plot of a ^{13}C T_1 Inversion- Recovery experiment

3.4 2D Experiments

3.4.1 $^1H/^1H$ Experiments

3.4.1.1 $^1H/^1H$ COSY Experiment [3.10, 3.11]

Theory

The $^1H/^1H$ COSY experiment is probably the most popular 2D experiment; it is used to correlate the chemical shifts of ^1H nuclei which are J-coupled to one another and to establish the $^1H/^1H$ coupling network (J-connectivity) of a molecule in a single experiment. There are many variations of the basic COSY experiment designed for specific applications such as the basic magnitude mode COSY (Fig. 3.20 top) for the rapid evaluation of coupling networks, the phase sensitive, double quantum (DQ-) filtered COSY (Fig. 3.20 bottom) for the detection of coupling networks and the measurement of the corresponding coupling constants, or the COSY experiment with selective presaturation for suppressing strong unwanted solvent signals and many others.

A first pulse creates transverse magnetization components (coherences) which evolve in the evolution period t1 (D0 in the schemes) with their characteristic precession frequencies (chemical shift and homonuclear J-coupling). The effect of the second (mixing) pulse is that information from one nucleus that evolves in t1 is transferred to another (J-coupled) nucleus, the magentization components of which evolve and are detected in t2. Therefore, the nuclei carry information that relates not only to their own chemical shifts and coupling constants but also the corresponding information about the other, coupled spins.

The COSY spectrum is produced by a double Fourier transformation with respect to t1 and t2, and its cross peaks indicate which ^1H nuclei are mutually J-coupled.

In its basic (magnitude mode) version the length of P0 is adjusted either to maximize the sensitivity (P0 = 90°) or to yield structured cross peaks (P0 = 45°). In the latter case information regarding the relative signs of coupling constants may be deduced.

The phase sensitive DQ-filtered COSY experiment has several significant advantages compared to the basic magnitude variant. It yields spectra with pure absorption lineshapes for the cross peaks (and the diagonal peaks) in F1 and F2. The coupling which causes the cross peak to appear, the active coupling, gives individual lines that are out-of-phase or in antiphase to each other, while the residual passive couplings give multiplet lines that are in-phase. Thus J-coupled connectivities and J values may be obtained from this type of experiment. To allow an accurate measurement of J values, the digital resolution is usually higher (at least in F2) compared to the basic magnitude mode COSY spectrum and consequently the measuring times are correspondingly longer. Diagonal peaks are partially cancelled which means that the diagonal ridge is much less pronounced than in a normal COSY spectrum and makes it easier to observe cross peaks between signals which are close together in chemical shift.

The double quantum filter eliminates or at least suppresses the strong signals from protons that do not experience J-coupling, e.g. the solvent signal, which would otherwise dominate the spectrum and possibly be a source of troublesome t1 noise. Compared to a phase-sensitive but non-DQ-filtered COSY with pure absorption lineshapes for the cross peaks but mixed lineshapes for the diagonal peaks, the phase-sensitive, DQ-filtered COSY has pure absorption lineshapes throughout.

Processing of a phase sensitve COSY spectrum, however, is complicated by the phase adjustments in both dimensions (see chapter 5).

The spectral quality and the efficiency of the basic COSY and the DQ-filtered COSY experiments may be improved with the use of field gradients instead of phase cycling for coherence selection, which remove spectral artifacts and make time consuming phase-cycling superfluous.

Pulse Diagrams

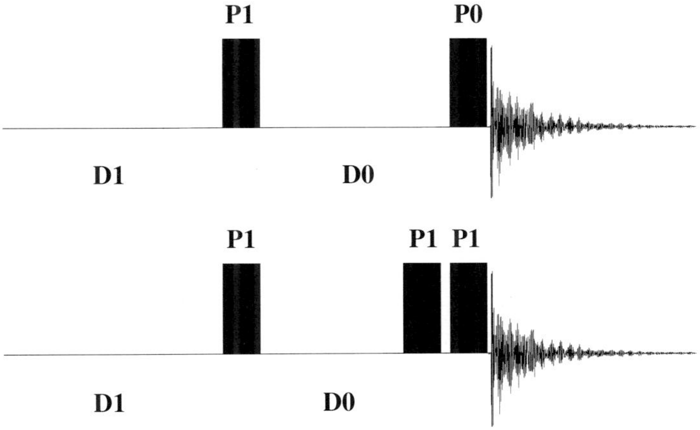

Fig. 3.20: The 2D COSY sequence: Top – Basic COSY experiment. Bottom – DQ-filtered COSY experiment

Application

The experiment is mainly used to establish the $^{1}H/^{1}H$ J-coupling network and to help assign the proton resonances of a molecule. Additional information, i.e. the evaluation of coupling constants can be obtained if the phase sensitive DQ-filtered COSY is used.

Example

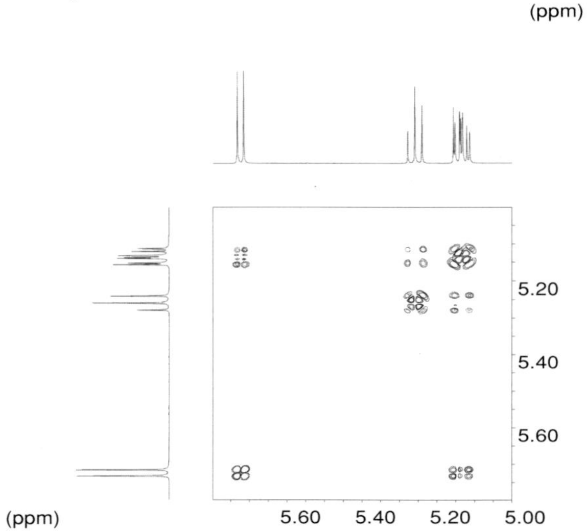

Fig. 3.21: 2D spectrum of a phase-sensitive, DQ-filtered COSY experiment

3.4.1.2 ^{1}H/^{1}H TOCSY Experiment [3.12, 3.13]

Theory

In contrast to COSY, TOtal Correlation SpectroscopY (TOCSY) uses cross polarization for coherence transfer in liquids as already discussed for the 1D TOCSY experiment (section 3.3.1.3). In 2D TOCSY, cross peaks are generated between all members of a coupled spin network. The experiment starts with a first evolution period t1 (D0 in Fig. 3.22), during which the coherences of a spin first evolve. The chemical shift and coupling information is transferred and distributed in the course of the spin-lock period – assuming the spin-lock period is long enough – in an oscillatory way among all the other coupled spins in the network. The length of the spin-lock period determines how "far" the spin coupling network will be probed. The coherences of these coupled spins are finally detected in t2 and carry information relating to their own chemical shifts and coupling constants as well the corresponding information about the spins in the same J-coupled spin system.

An advantage of 2D TOCSY is that the "net" coherence transfer produced can be arranged to create pure positive absorption spectra, including the diagonal peaks, rather than spectra with equal positive and negative intensities obtained with "differential" coherence transfer as in the COSY experiment.

Pulse Diagram

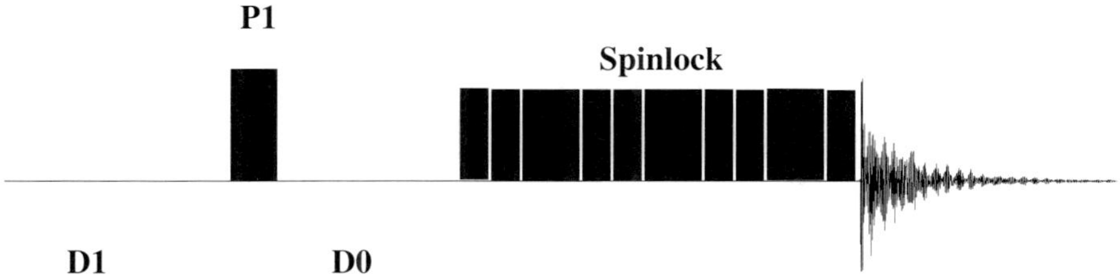

Fig. 3.22: The 2D TOCSY pulse sequence

Application

The experiment is used to identify the subspectra of isolated spin systems and is most often used for molecules which consist of rows or networks of similar fragments with no coupling interactions between them, e.g. oligosaccharides, oligopeptides or oligonucleotides. The experiment is usually combined with a COSY experiment, where it helps to overcome problems of overlapping cross peaks that can arise in the latter case.

Example

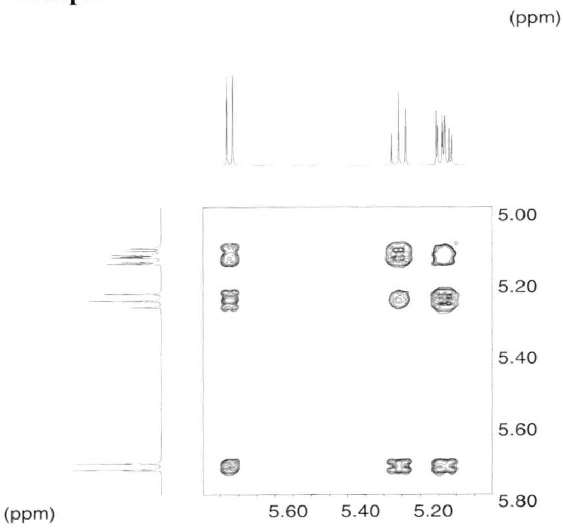

Fig. 3.23: The 2D spectrum from a 2D TOCSY experiment. The same sample has been used and the expansion is the same as for the 2D phase-sensitive, DQ-filtered COSY spectrum (Fig. 3.21). Note the additional cross peaks obtained with the TOCSY experiment.

3.4.1.3 ^{1}H/^{1}H NOESY and ^{1}H/^{1}H ROESY Experiments [3.14, 3.15]

Theory

The basic NOESY (NOE SpectroscopY) sequence (Fig. 3.24, top) consists of three 90° pulses. The first pulse creates transverse spin magnetization (coherence). This precesses during the evolution time t1 (D0 in the scheme). The second pulse produces longitudinal magnetization equal to one of the transverse magnetization components (x, y). Thus the basic idea is to produce an initial situation for the mixing period D9 (the time during which cross relaxation occurs) where the longitudinal magnetization of each spin is labelled by its chemical shift. The longitudinal magnetization is allowed to relax and NOEs are built up for other nuclei close in space during the mixing time D9. Therefore the NOE transferred to these other spins is modulated in t1 and the modulation frequency corresponds to the chemical shift of the nuclei responsible for the NOE. This information is probed by the third pulse creating transverse magnetization which is detected in t2.

Rotating frame Overhauser Effect SpectroscopY (ROESY) is an experiment in which homonuclear NOE effects are measured under spin-locked conditions as outlined in detail for the 1D ROESY experiment (section 3.3.1.5). The experiment (Fig. 3.24, bottom) starts with a 90° pulse prior to the evolution period t1 during which the same situation as in the NOESY experiment is produced for the subsequent spin-lock period. In contrast to the NOESY experiment, where one of the transverse magnetization components (x, y) is converted into longitudinal magnetization prior to cross relaxation, one of these transverse components is spin-locked and cross relaxation occurs under spin-lock conditions. The size of this transverse magnetization is modulated in t1 with the chemical shift frequency of the corresponding spin. At the same time, and after an ROE has built up in the spin-lock period for all spins closely related in space, this frequency modulates the intensities of their signals which are finally detected in t2.

NOESY and ROESY spectra are usually measured in phase sensitive mode. The cross peaks in a NOESY spectrum indicate spatial proximity between the protons that give rise to the corresponding diagonal peaks. Depending on molecular size and solvent viscosity the cross peaks may display negative absorptive (small highly mobile molecules), or positive absorptive (large, slowly tumbling molecules) with respect to the positive absorptive diagonal peaks. For ROESY spectra, however, cross peaks and diagonal peaks show absorption lineshapes of opposite sign, irrespective of the size of the molecule under investigation. This makes ROESY experiments more suitable for molecules of intermediate size, where NOEs may be close or equal to zero, as discussed for the 1D NOE experiment (section 3.3.1.4).

In contrast to the 1D experiment, where "steady-state" NOEs may be obtained, only the less intense transient NOEs are measured in the NOESY experiment. ROEs can only be obtained as transient effects in both the 1D and the 2D experiment. Furthermore the intensities of the NOESY and ROESY cross peaks depend upon the molecular size as well as the length of the mixing period. In the case of large molecules, e.g. polypeptides, rather short mixing times are usually chosen to avoid spin diffusion.

Occasionally, COSY-type artifacts appear in NOESY and ROESY spectra but these are easy to identify by their anti-phase multiplet structure.

In the case of chemical or dynamical exchange processes, cross peaks originating from saturation transfer are superimposed on the normal NOESY and ROESY cross peaks in the spectrum. For small molecules they may easily be identified since they appear in-phase with respect to the diagonal peaks and are in most cases rather intense.

Pulse Diagrams:

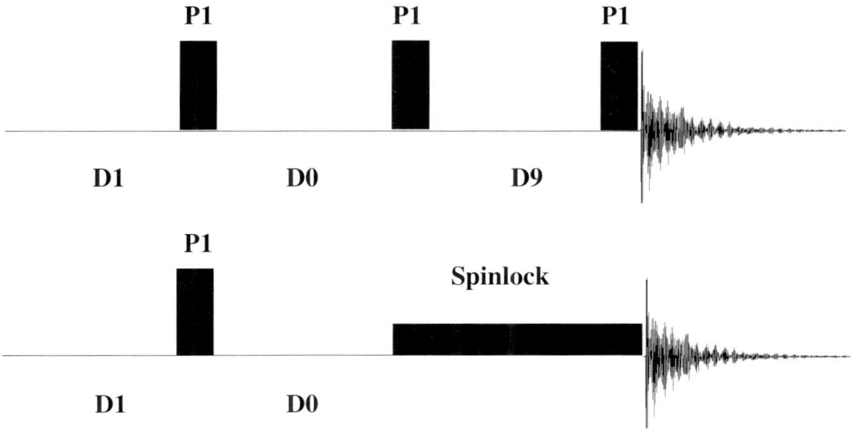

Fig. 3.24: The 2D NOESY (top) and the 2D ROESY (bottom) pulse sequences.

Application

The experiments are used to recognise the spatial proximity among the protons of a molecule. The experiments are useful if a large number or all proton-proton interactions within a molecule or between molecules are of interest. If the spatial arrangement of only a few protons are of interest, a common situation with small molecules, the 1D NOE experiment with its inherent higher sensitivity offers a better alternative.

The 2D NOESY and the 2D ROESY experiments may also be used to measure NOE- or ROE- build-up rates. This is accomplished using a series of experiments, where the mixing period D9 or the length of the spin-lock period respectively is incremented from experiment to experiment. From build-up rates relative internuclear distances may be estimated and calculated.

A further field of application for the 2D NOESY and the 2D ROESY experiments are “dynamic” systems, where exchange processes may be recognized and may be analysed quantitatively (EXCSY-spectroscopy).

Example

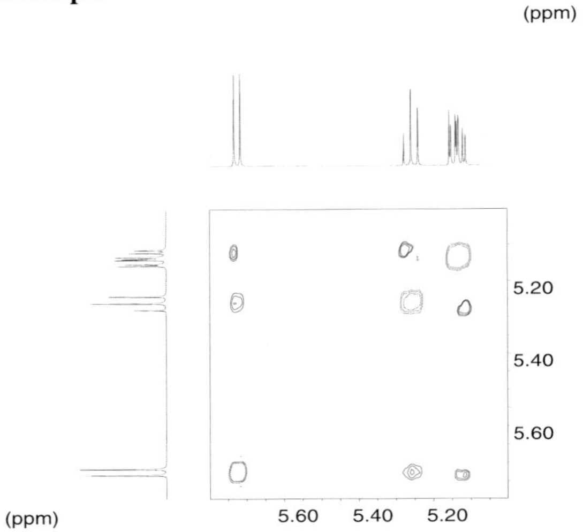

Fig. 3.25: The 2D spectrum of a 2D ROESY experiment. Both the positive cross peaks and the negative diagonal peaks are shown.

3.4.1.4 ^{1}H/^{1}H J-Resolved Spectroscopy Experiment [3.16]

Theory

2D J-Resolved spectroscopy is an experiment to separate J-coupling from chemical shifts and may be performed in a homonuclear or heteronuclear mode. The homonuclear experiment (Fig. 3.26) starts with a 90° pulse generating transverse magnetization, which evolves in t1 (D0 in the scheme). The 180° pulse centered in t1 refocusses the effect of chemical shifts but not the effect of homonuclear J-coupling (spin-echo). Therefore J-information is probed in t1 and both J- and chemical shift information is probed in t2. 2D Fourier transformation yields a spectrum with J-couplings appearing in F1 and J-couplings/chemical shifts appearing in F2. A subsequent processing step (tilting the 2D matrix), "removes" the J-coupling in F2 and allows the separation of the two parameters in F1 and F2.

Pulse Diagram

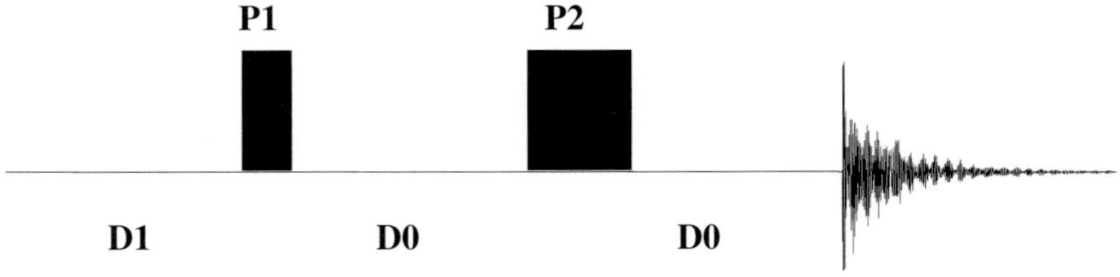

Fig. 3.26: The homonuclear 2D J-Resolved pulse sequence

Application

The experiment is used to separate chemical shifts and J-couplings for homo- and heteronuclear systems. In simple cases the chemical shifts and J-couplings may be directly obtained from the 2D spectrum by inspection. For severely overlapped first-order spectra or strongly coupled spin systems the estimated parameters obtained from the spectrum may be used as starting values in a computer assisted spectral analysis as outlined in *Modern Spectral Analysis* (Volume 3).

Example

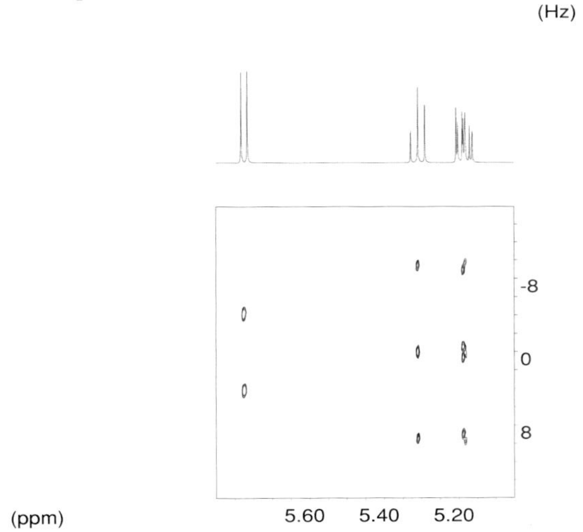

Fig. 3.27: 2D spectrum (tilted) from a homonuclear 2D J-Resolved experiment.

3.4.2 $^{1}H/^{13}C$ Experiments

3.4.2.1 $^{1}H/^{13}C$ Shift Correlation Spectroscopy via $^{1}J_{CH}$ [3.17 - 3.21]

Theory

Heteronuclear shift correlation spectroscopy is a 2D technique that can be used to determine the connectivity of ^{1}H and ^{13}C nuclei (or other heteronuclei), formally bonded together through one or more chemical bonds. The corresponding experiments make use of either the large $^{1}J_{XH}$ or the smaller long-range $^{n}J_{XH}$ couplings for polarization transfer.

A variety of sequences exist, which differ with respect to the detected interaction ($^{1}J_{XH}$ or $^{n}J_{XH}$) and the mode of detection (^{13}C or ^{1}H detected, magnitude or phased mode, phase cycling or gradients for coherence selection). In view of the reduced sensitivity of heteronuclear experiments with respect to homonuclear COSY experiments and the steadily decreasing sample amounts submitted for NMR experiments, there is no doubt that the inverse (^{1}H) detected, gradient enhanced experiments are currently the best methods to apply. However on older type spectrometers, not equipped for inverse detection the "old-fashioned" direct ^{13}C detected experiments are still in use.

The pulse schemes and corresponding 2D spectra for three typical and popular representatives are shown in Figs. 3.28 and 3.29 respectively.

The pulse sequence for the ^{13}C detected 2D $^{13}C/^{1}H$-COSY experiment to measure $^{1}J_{CH}$ connectivities (Fig. 3.28a) starts, after a preparation period D1, with a 90° proton pulse generating transverse proton magnetization, which evolves in t1 (D0 in the scheme) under the influence of ^{1}H chemical shifts and $^{n}J_{HH}$ couplings. This ^{1}H information is transferred using two 90° polarization transfer pulses, applied simultaneously to both the ^{1}H and ^{13}C channels, to the directly bonded ^{1}J-coupled ^{13}C nucleus and yields a series of amplitude modulated ^{13}C spectra where the modulation frequencies carry the ^{1}H chemical shift and $^{n}J_{HH}$ coupling information. Heteronuclear couplings are "removed" in both dimensions by adding a ^{13}C 180° pulse in the middle of t1 and by ^{1}H broadband decoupling during data acquisition.

The final 2D spectrum has a projection onto the F2 axis which is the normal ^{1}H-decoupled ^{13}C spectrum with the quaternary carbons missing, and a projection onto the F1 axis which is the normal ^{1}H spectrum, including $^{n}J_{HH}$ coupling. This experiment is not phase-sensitive, and must be displayed in magnitude mode.

To enhance the sensitivity of the experiment and to simplify the spectra a BIRD filter (not shown in the scheme) is usually introduced in the middle of the evolution period t1. With the exception of geminal $^{2}J_{HH}$ couplings, this cancels the effect of all residual $^{n}J_{HH}$ couplings and yields a projection onto the F1 axis displaying the ^{1}H chemical shifts alone.

Two variations of the most popular ^{1}H detected 2D $^{13}C/^{1}H$-COSY experiment to measure $^{1}J_{CH}$ connectivities (Fig. 3.28b and 3.28c) exist, differing in the spin states exploited for shift correlation using $^{1}J_{CH}$ and the evolution in t1. The HMQC (Heteronuclear Multiple Quantum Coherence) experiment uses multiple quantum coherence, while the HSQC (Heteronuclear Single Quantum Coherence) experiment uses single quantum coherence for this purpose. The advantage of both sequences is their inherent higher sensitivity compared to the direct ^{13}C detected 2D $^{13}C/^{1}H$-COSY experiment. The challenge of inverse shift correlation experiments, however, is the suppression of the large unwanted signals induced by protons not directly bound to a ^{13}C nucleus and originating almost exclusively from the all ^{12}C isotopomers. This is accomplished using either multiple quantum filters (HMQC), adequate phase cycling and/or magnetic field gradients.

In the HMQC sequence (Fig. 3.28b), the first ^{1}H pulse creates transverse magnetization, which evolves into anti-phase magnetization during the first D2 delay, adjusted to $1/(2\ ^{1}J_{XH})$. This anti-phase magnetization is converted into zero- and double quantum coherence by the first 90° ^{13}C pulse and evolves in t1 (D0 in the scheme). Due to the 180° ^{1}H pulse centered in t1, only single quantum ^{13}C frequencies, i.e. the ^{13}C chemical shifts, superimposed with $^{n}J_{HH}$ are monitored in t1. The final 90° ^{13}C pulse converts multiple quantum coherence into observable ^{1}H transverse magnetization, the amplitude of which is modulated by the ^{13}C chemical shift information. In analogy with the basic ^{13}C detected experiment, the final delay D2 refocusses anti-phase magnetization prior to the acquisition with broadband ^{13}C decoupling of the ^{1}H signals.

In the HSQC sequence (Fig. 3.28c), the first ^{1}H pulse creates transverse magnetization, which is refocussed to anti-phase magnetization prior to the polarization transfer. Polarization transfer is accomplished in an INEPT style using 90° pulses applied simultaneously to both the ^{1}H and ^{13}C channels. The resulting ^{13}C single quantum coherence evolves in t1 (D0 in the scheme) and the simultaneously applied 90° ^{1}H and 90° ^{13}C pulses transfer the ^{13}C anti-phase coherence into ^{1}H anti-phase coherence. Due to the 180° ^{1}H pulse centered in t1, single quantum ^{13}C frequencies, i.e. the ^{13}C chemical shifts, are monitored in t1. In contrast to the HMQC experiment no additional $^{n}J_{HH}$ couplings are superimposed on the signals. The final pulses and delays transfer this information onto the coupled proton (retro-INEPT step) and the ^{1}H signals are acquired with broadband ^{13}C decoupling.

In contrast to the basic ^{13}C detected experiment, and as a consequence of the final ^{1}H detection, the 2D spectra obtained with HMQC or HSQC have a projection onto the F2 axis which corresponds to the normal ^{1}H spectrum and includes all chemical shifts and all J_{HH} couplings. The latter may give rise to rather broad cross peaks for extensively coupled protons. The projection onto the F1 axis corresponds to a normal ^{13}C spectrum but with the quaternary carbons missing. With HMQC, but not with HSQC, cross peaks are additionally split in F1 by $^{n}J_{HH}$ couplings. The HMQC and the HSQC experiment are usually performed in phase-sensitive mode, which, after proper phasing in both dimensions, allow peaks to be displayed in pure absorption.

To improve the spectral quality the unwanted ^{1}H signals of the ^{12}C isotopomers may be further reduced using either a BIRD pulse during the preparation period or replacing the phase cycling by field gradients which will allow the proper selection of the coherence pathways in conjunction with the HMQC and HSQC experiment. Of the two methods for suppressing the unwanted ^{12}C signals, the gradient method is the more efficient. With both experiments the same detection sensitivity is obtained. A gradient-enhanced version of the HSQC experiment has recently been developed which is based upon and takes advantage of the echo-antiecho mode of detection. Using this sequence the detection sensitivity is enhanced by a factor of 2 compared to the normal experiment.

Pulse Diagrams

a)

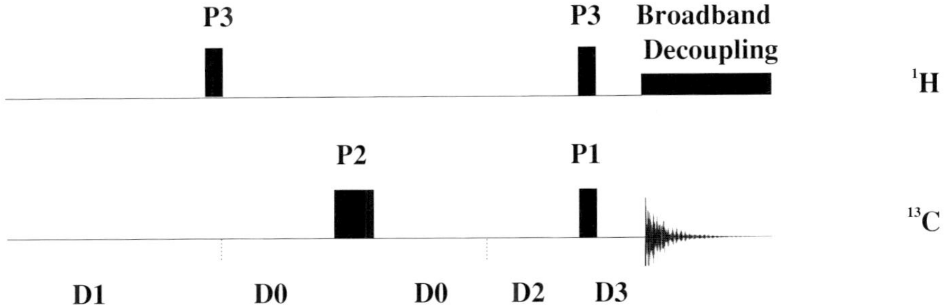

Fig. 3.28: The $^{13}C/^{1}H$ COSY experiments: ^{13}C-detected $^{13}C/^{1}H$-COSY (a)

b)

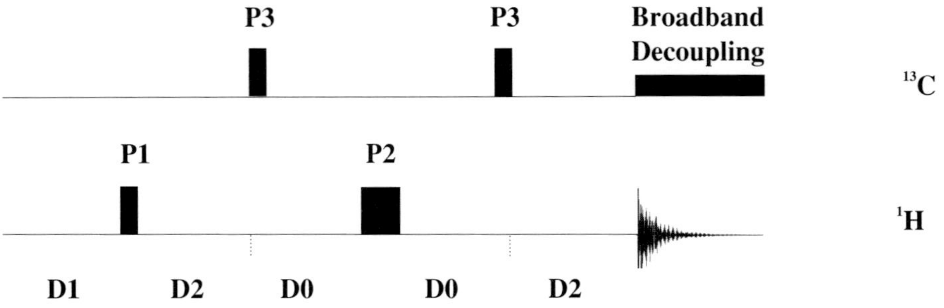

c)

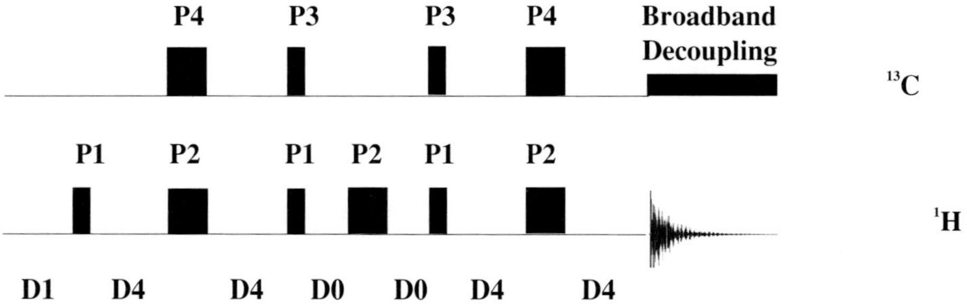

Fig. 3.28 (continued): The $^{13}C/^{1}H$-COSY experiments: ^{1}H-detected HMQC (b), ^{1}H-detected HSQC (c).

Application

The experiments are used to correlate ^{1}H- and ^{13}C chemical shifts. The ^{13}C detected $^{13}C/^{1}H$-COSY experiment (Fig. 3.29a) is the most popular heteronuclear 2D experiment available with older type spectrometers, not yet equipped with "inverse" ^{1}H detection mode. Due to the surviving geminal couplings in F1 the signals of methylene groups with non-equivalent protons are immediately recognized.

Due to their higher sensitivity, the ^{1}H detected $^{13}C/^{1}H$-COSY experiments HMQC (Fig. 3.29b) and HSQC (not shown) are more suitable if only small sample amounts are available.

With long acquisition times and a correspondingly high resolution in F2, the experiment may be particularly useful in cases where the ^{1}H spectrum is overcrowded. The corresponding ^{13}C spectra are usually well resolved and the separation of the cross peaks in F1 of such 2D experiments can be used to extract the corresponding ^{1}H sub-spectra (rows of the 2D matrix) which may then be analysed separately and allow to obtain ^{1}H chemical shifts and $^{1}H/^{1}H$ coupling constants even for such demanding cases.

Examples

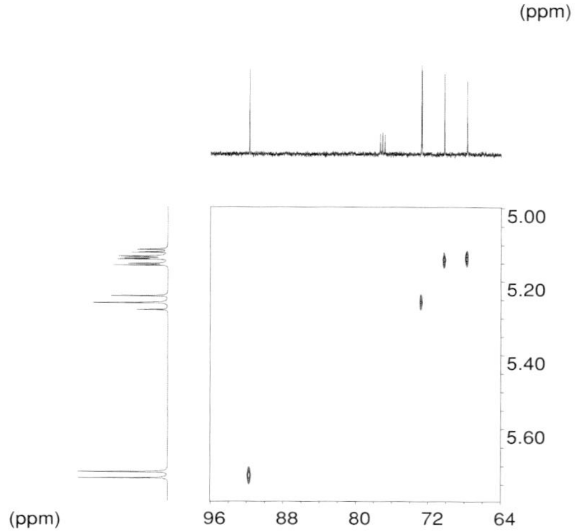

Fig. 3.29a: 2D spectrum of the ^{13}C-detected, magnitude mode ^{13}C/^{1}H-COSY experiment

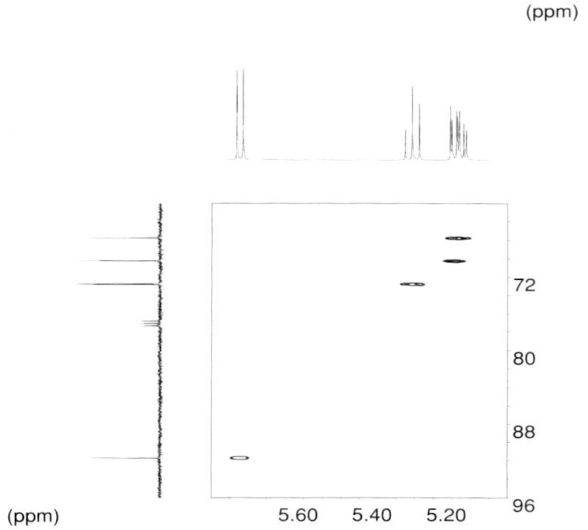

Fig. 3.29b: 2D spectrum of the ^{1}H-detected, phase-sensitive HMQC experiment

3.4.2.2 ^{1}H/^{13}C Shift Correlation Spectroscopy via $^{n}J_{CH}$ [3.22]

Theory

Heteronuclear (X, H) Multiple Bond Correlation (HMBC) spectroscopy is a modified version of the HMQC experiment and is suitable for determining long-range ^{1}H/^{13}C connectivities. This is useful for signal assignments, including quaternary carbons, and structure elucidation. HMBC provides basically the same information as the popular COLOC experiment but takes advantage of the higher sensitivity of inverse ^{1}H detection.

The HMBC pulse sequence (Fig.3.30) starts with a 90° ^{1}H pulse creating transverse magnetizations of protons coupled directly ($^{1}J_{CH}$) and via several bonds ($^{n}J_{CH}$) to the ^{13}C in the molecule. The first 90° ^{13}C pulse acts as a low-pass filter and transfers the former into heteronuclear multiple quantum coherence; the cross peaks generated by $^{1}J_{CH}$ are then suppressed in the final 2D spectrum by appropriate phase cycling. Transverse magnetization of protons coupled via several bonds ($^{n}J_{CH}$) is transferred into double quantum coherence with the second 90° ^{13}C pulse. Due to the 180° ^{1}H pulse centered in t1, only single quantum ^{13}C frequencies are monitored in t1. The final 90° ^{13}C pulse converts multiple quantum coherence into observable ^{1}H transverse magnetization, the amplitude of which is modulated by the ^{13}C chemical shift and the homonuclear coupling ($^{n}J_{HH}$) information. No broadband ^{13}C decoupling is applied during acquisition.

The 2D spectrum (Fig. 3.31) has a projection onto the F2 axis which corresponds to the normal ^{1}H spectrum, displaying all chemical shifts, all $^{n}J_{HH}$ and $^{n}J_{CH}$ couplings. The $^{n}J_{HH}$ coupling may again give rise to rather broad cross peaks for extensively coupled protons. The projection onto the F1 axis gives a ^{13}C spectrum with the chemical shifts and the J_{HH} couplings of the corresponding directly bonded protons.

To improve the spectral quality the unwanted ^{1}H signals of the ^{12}C isotopomers may be further reduced by using field gradients which will allow the proper selection of the coherence pathways in conjunction with the HMBC experiment

The cross peaks in the 2D spectrum are a combination of absorption and dispersion lineshapes and consequently spectra are displayed in magnitude mode.

Pulse Diagram

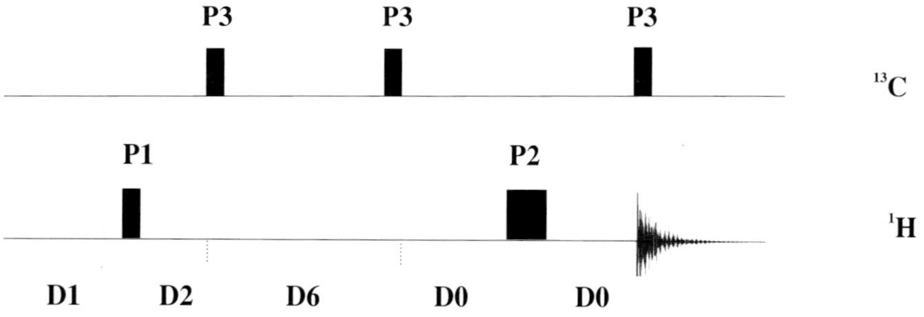

Fig. 3.30: The HMBC pulse sequence

Application

The experiment is applied to correlate ^{1}H and ^{13}C chemical shifts of nuclei coupled through a number of bonds using $^{n}J_{CH}$. This experiment is superior to the popular COLOC experiment due to its higher sensitivity and is therefore recommended for small sample amounts.

It enables the assigment of quaternary carbons, which often act as barriers in the process of structure elucidation, and is used to connect molecular fragments, the structure of which have already been determined using other information.

Example

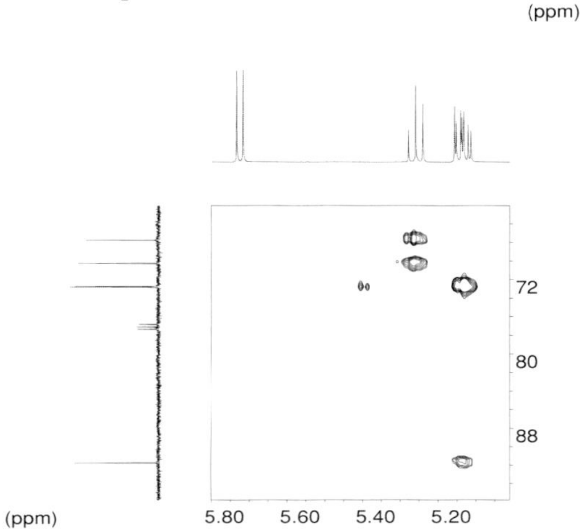

Fig. 3.31: 2D spectrum of the HMBC experiment

3.5 Recommended Reading

Introductory Textbooks and Reviews:

Croasmun, W. R., Carlson, R. M. K., *Two-Dimensional NMR Spectroscopy:Applications for Chemists and Biochemists*, VCH, 1994

Derome, A. E., *Modern NMR Techniques for Chemistry Research*, Pergamon Press, Oxford, 1987

Friebolin, H. P., *Basic One- and Two-Dimensional NMR Spectroscopy*, VCH, 1993

Kessler, H., Gehrke, M., Griesinger, C., *Two-Dimensional NMR Spectroscopy: Background and Overview of the Experiments*, Angew. Chem. Int. Ed. Engl., 1988, *27*, 490

Martin, G. E., Zektzer, *A. S., Two-Dimensional NMR Methods for Establishing Molecular Connectivity*, VCH, Weinheim, 1988

Ziessow, D., *Understanding Multiple-Pulse Experiments*, Concepts in Magnetic Resonance; An Educational Journal, 1990, *2* (No 2)

Rahman, A. U., *One and Two Dimensional NMR Spectroscopy*, Elsevier, 1989

Sanders, J. K. M., Hunter, B. K., *Modern NMR Spectroscopy; A Guide for Chemists*, Oxford University Press, 1993

Schraml, J., Bellama, M., *Two Dimensional NMR Spectroscopy: A Textbook for Chemists*, Wiley, 1988

More Special Textbooks, Reviews and References:

Braun, S., Kalinowski, H.-O., Berger, S., *100 and More Basic NMR Experiments; A Practical Course*, VCH, 1996

Ernst, R. R., Bodenhausen, G. Wokaun, A., *Principles of Nuclear Magnetic Resonance in One and Two Dimensions*, Clarendon Press, Oxford, 1990

Farrar, T. C., *Density Matrices in NMR Spectroscopy*, Concepts in Magnetic Resonance; An Educational Journal, 1990, *2* (No 1, 2)

Frye, J. S., *Comparison of Inversion-Recovery Methods for Measuring Longitudinal Relaxation Rates*, Concepts in Magnetic Resonance; An Educational Journal, 1989, *1* (No 1)

Kingsley, P.B., *Product Operators, Coherence Pathways and Phase Cycling*, Concepts in Magnetic Resonance; An Educational Journal, 1995, 7 (No 1, 2 and 3)

Lowry, D. F., *Correlated Vector Model of Multiple-Spin Systems*, Concepts in Magnetic Resonance; An Educational Journal, 1994, *6* (No 1)

Martin, M. L., Delpuech, J-J., Martin, G. J., *Practical NMR Spectroscopy*, Heyden, 1980

Popov, A. I., Hallenga, K., *Modern NMR Techniques and their Application in Chemistry*, Marcel Dekker , published in USA, 1990

Shriver, J., *Product Operators and Coherence Transfer in Multiple-Pulse NMR Experiments*, Concepts in Magnetic Resonance; An Educational Journal, 1992, *4* (No 1)

Sørensen, O. W., Eich, G. W., Levitt, M. H., Bodenhausen, G., Ernst, R. R., *Product Operator Formalism for the Description of NMR Pulse Experiments*, Prog. Nucl. Magn. Reson. Spectrosc., 1983, *16*, 163

Wemmer, D. E., *Homonuclear Correlated Spectroscopy (COSY)*, Concepts in Magnetic Resonance; An Educational Journal, 1989, *1,* (No 2)

Zhu, J. M., Smith, I. C. P., *Selection of Coherence Transfer Pathways by pulsed-Field Gradients in NMR Spectroscopy*, Concepts in Magnetic Resonance; An Educational Journal, 1995, *7* (No 4)

References to the Experiments:

[3.1] Philipsborn, W. v., *Angew. Chem.* internat. Edit., 1971, *83* (No 13), 472
[3.2] Davis, D. G., Bax, A., *J. Amer. Chem. Soc.*, 1985, *107*, 7197
[3.3] Neuhaus, D., Williamson, M. P., *The Nuclear Overhauser Effect in Structural and Conformational Analysis*, VCH, N.Y., 1989
[3.4] Bothner-By, A. A., Stephens, R. L. J., Lee, Warren, C. D., Jeanloz, R.W. J., *J. Amer. Chem. Soc.*, 1984, *106*, 811
[3.5] Doddrell, D. M. Pegg, D. T., Bendall, M. R., *J. Magn. Reson.*, 1982, *48*, 323
[3.6] Le Cocq, C., Lallemand, J.-Y., *J. Chem. Soc. chem. commun.*, 1981, 150
[3.7] Patt, S. L., Shoolery J. N., *J. Magn. Reson.*, 982 *46*, 535
[3.8] Vold, R. L., Waugh, J. S., Klein, M. P., Phelps, D. E., *J. Chem. Phys.*, 1968, *48*, 3831
[3.9] Freeman, R., Hill, H. D., *J. Chem. Phys.*, 1969, *51*, 3140
[3.10] Aue, W. P., Bartholdi, E., Ernst, R. R., *J. Chem. Phys.*, 1976, *64*, 2229
[3.11] Wokaun, A., Ernst, R. R., *Chem. Phys. Lett.*, 1977, *52*, 407
[3.12] Braunschweiler, L., Ernst, R. R., *J. Magn. Reson.*, 1983, *53*, 521
[3.13] Bax, A. Davis, D. G. *J. Magn. Reson.*, 1985, *65*, 355
[3.14] Bodenhausen, G., Kogel, H., Ernst, R. R., *J. Magn. Reson.*, 1983, *58*, 370
[3.15] Hwang, T. L., Shaka, A., *J. Amer. Chem. Soc.*, 1992, *114*, 3157
[3.16] Aue, W. P., Karhan, J., Ernst, R. R., *J. Chem . Phys.*, 1976, *65*, 4226
[3.17] Freeman, R., Morris, G. A., *J. Chem. Soc. Chem. Commun.*, 1978, 684
[3.18] Bax, A., *J. Magn. Reson.*, 1983, *53* 517
[3.19] Bax, A., Griffey, H., Hawkins, B. L., *J. Magn. Reson.*, 1983, *55* 301
[3.20] Bodenhausen, G., Ruben, D. J., *Chem. Phys. Lett.*, 1980, 69, 185
[3.21] Palmer, A. G., Cavanagh, J., Wright, P. E., Rance, M., J., *J. Magn. Reson.*, 1991, 93, 151
[3.22] Bax, A., Summers, M. F., *J. Amer. Chem. Soc.*, 1986, *108*, 2093

4 How to Display and Plot 1D and 2D NMR Spectra

4.1 Introduction

This chapter describes the basic features of the 1D and 2D WIN-NMR software for displaying and plotting 1D and 2D NMR spectra and includes simple data manipulation (or final processing) prior to plotting. This introduction will allow you to display, inspect and plot any of the spectra in the NMR data base in a standard way or according to your needs and preferences. An overview of the various options available with the WIN-NMR software is given. For greater detail refer to the information available with the **Help** options together with the corresponding *Check its*. This will allow you to become familiar with the many features of these processing software tools.

In many cases you will have access to NMR data already pre-processed either by an NMR specialist or from a spectrometer working in automation mode. Therefore displaying and plotting spectra will be the most frequent tasks associated with the processing of NMR data that you will undertake during the course of your laboratory work. Plotting spectra, however, is not always a necessity, since for controlling the progress of synthetic work in the laboratory, or for solving simple structural problems the display and inspection of the spectrum on the screen may be sufficient and less time-consuming.

Using the spectra supplied in the NMR data base you will become proficient in the displaying, final processing and plotting of 1D and 2D spectra. You will then be ready to progress onto Chapter 5, where the NMR raw data (FID) rather than the pre-processed spectra is the starting-point for processing, and where the many processing options and their effects on the final spectra will be discussed.

It is assumed at this stage that the 1D and 2D data to be displayed and plotted have already been processed and that the spectra are available with all the signals correctly phased. Processing such as weighting the FID, Fourier transformation and phasing the spectrum will be discussed in chapter 5. Before processed NMR spectra can be displayed and plotted correctly a few final processing steps such as peak calibration, peak picking and peak integration must be completed.

The main steps for displaying and plotting NMR spectra are shown in Fig. 4.1.

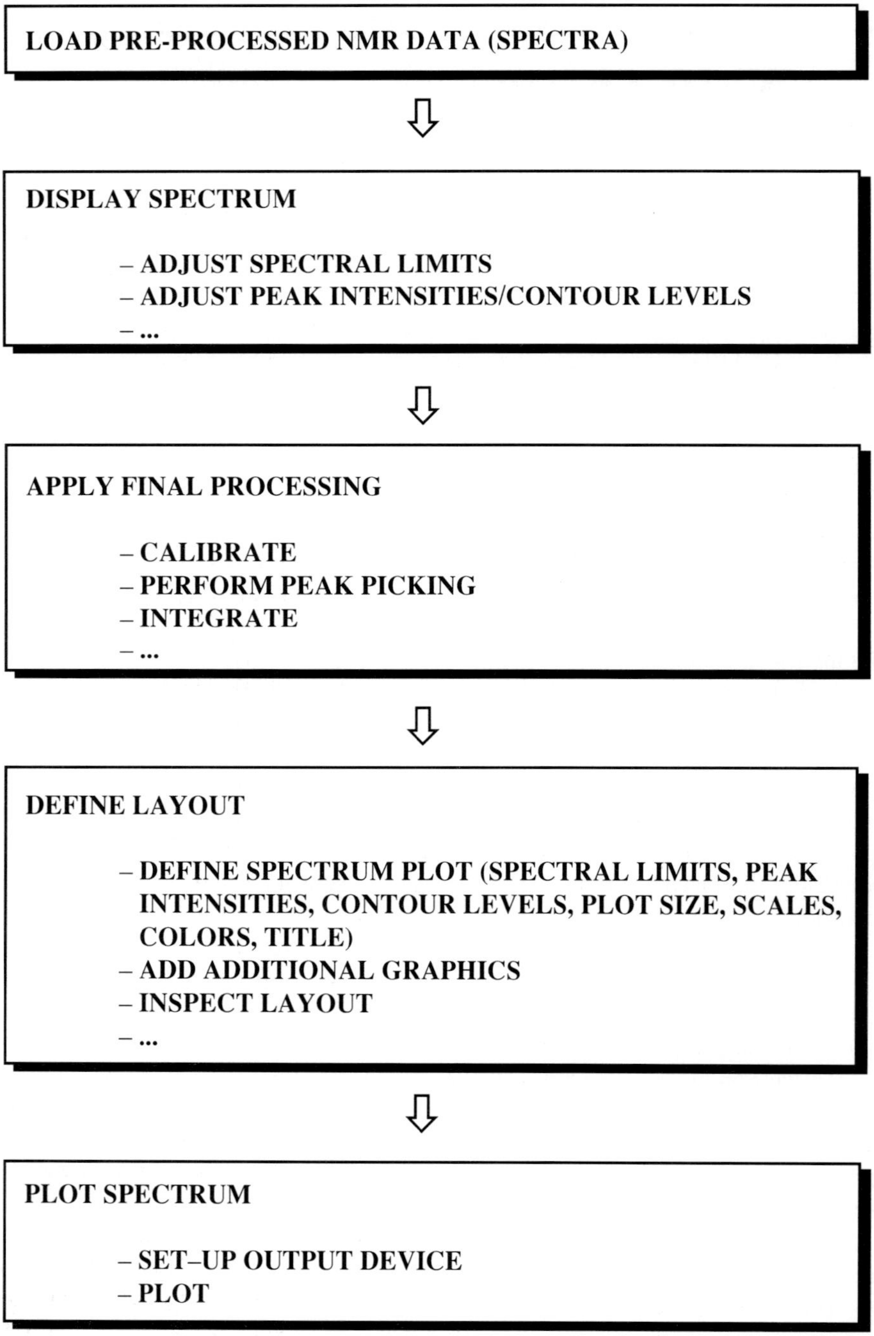

Fig. 4.1: The main steps for displaying and plotting NMR spectra.

To set up the page layout for plotting spectra several tasks have to be performed. They include the definition of the plot region, of the spectrum components to be included in the plot (spectrum, integral, peak labels, axis, ...), of the title, of the contents of the parameter table and any additional imported data to be added to the layout such as the formulae or molecular structure. To inspect the final layout prior to plotting, WIN-NMR offers the **Preview** option which shows an exact copy of your subsequent plot on the screen and allows you to perform final adjustments and rearrangements on the screen.

Final processing and arranging the page layout may be performed in different ways. Using the appropriate quick-mode buttons is the easiest way, whereas the use of pull-down menus offers a wide variety of parameters which allow the various operations to be tailored to your specific needs. In addition, some of these operations may be performed in an automatic, but predefined way, which is ideal if a series of spectra are to be processed and plotted in the same way (see section 5.5).

4.2 Help Routines

The 1D WIN-NMR and the 2D WIN-NMR **Help** options will give you detailed information on a given topic as well as how to apply the corresponding processing options. 1D WIN-NMR has seven and 2D WIN-NMR has ten main topics:

1D WIN-NMR:	**2D WIN-NMR:**
Introduction	General information
Initialization Files	Installation
Basic Topics	Buttons Panel
FID- and Spectrum Window	File
Preview Window	Simulation
Relaxation Window	Process
Text Window	Analysis
	Output
	Display
	Help

The **Help** pages available via the corresponding pull-down menu, are organized in a hierarchy with information on different levels. Two ways to exploit **Help** exist, either by using the **Help** menu system or by using a special cursor. Starting from a main topic associated subtopics on lower levels may be accessed.

If you start the **Help** routine via **Contents** (1D WIN-NMR) or **Help index** (2D WIN-NMR) and you click on one of the underscored topics in the **Help** menu, you will get a new **Help** page on a lower level containing more information about this topic. Self-explanatory buttons allow you to move up and down in the **Help** hierarchy and to search for the required information.

If you start the **Help** routine in 1D WIN-NMR via **?** the cursor shape changes and you can directly select with this new mouse cursor the appropriate information about any entry in the pull-down menus, or any button or function key in the display button panel.

Check it in 1D WIN-NMR:

Start the 1D WIN-NMR program and from the first **Help** menu choose the **Contents** submenu to access the introductory remarks and the main **Help** menu. Click on the item **FID- and Spectrum Window** and select the topic **Pull Down Menu**. Further items appear. Select **File** and then **Filecopy & Convert**. Here you will find all the necessary information for the transfer of UXNMR/XWINNMR files from a remote computer to your PC. Use the **Up** button to move up from one level to the other, back to the main **Help** menu and the **Close** button to exit the **Help** tool.

Check it in 1D WIN-NMR:

From the 1D WIN-NMR **Help** menu choose the entry **?**, which switches you back to the 1D WIN-NMR application window with a reshaped mouse pointer. From **File** in the main menu bar choose the **Filecopy & Convert** option. This will open the same information menu as above, but in more direct and convenient way.

Check it in 1D WIN-NMR and 2D WIN-NMR:

Select other items in the 1D WIN-NMR and 2D WIN-NMR **Help** menus and browse through the items of interest until you are familiar with using these tools.

4.3 Application Windows for 1D WIN-NMR and 2D WIN-NMR

Whenever 1D WIN-NMR or 2D WIN-NMR start, the appropriate application window appears on screen (see Figs. 2.5 and 2.6). This window may be minimized, maximized and resized using the standard MS-WINDOWS techniques. Access to the MS-WINDOWS system task menu is by clicking the button positioned to the left of the title bar. This menu may be used to switch to other running programs or to close one of the WIN-NMR sessions. Refer to the MS-WINDOWS manual for a detailed explanation of the individual window elements and for specific window manipulation techniques. As an example the 1D WIN-NMR application window is shown. It consists of several menus, buttons and subwindows (Fig. 4.2).

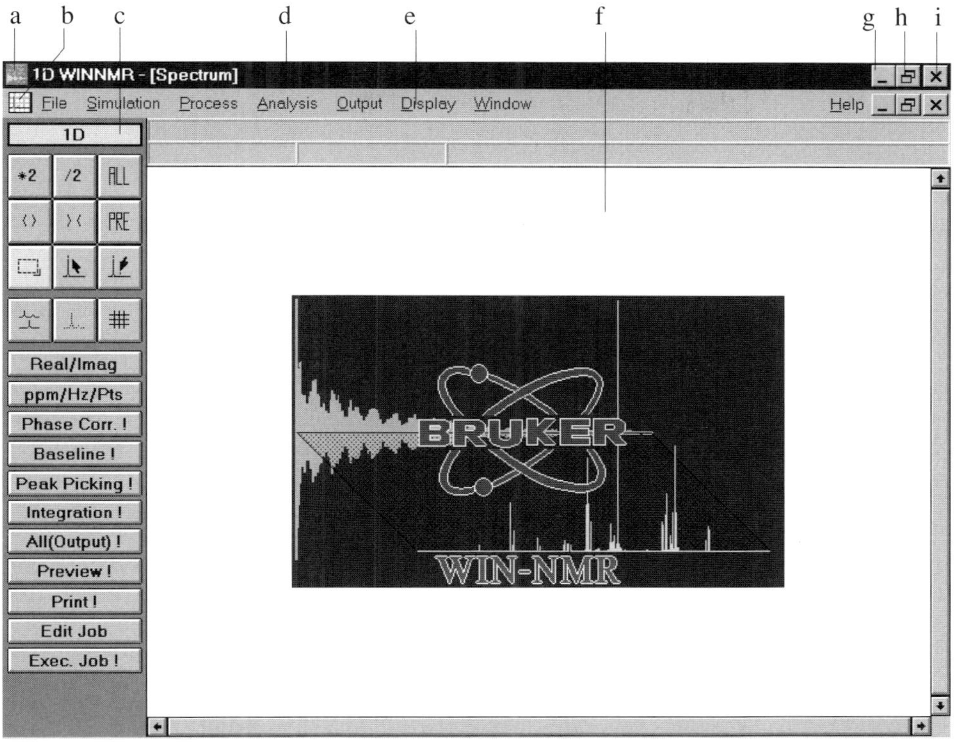

Fig. 4.2: 1D WIN-NMR application window after program start. a) system menu button, b) MDI (Multi-Document-Interface) system menu button, c) button panel, d) title bar, e) menu bar, f) *Spectrum* window, g) button to iconize the window h) minimize/maximize button, i) close button.

The items in the menu bars of 1D WIN-NMR and 2D WIN-NMR open a series of pull-down menus:

File This menu contains commands to open and store data files, to access the MS-WINDOWS Program Manager, to call the GETFILE program for copying and converting data files stored on remote computers, to extract FIDs from a 2D spectrum into 1D WIN-NMR spectra and to exit the WIN-NMR program.

Simulation Calls various simulation programs such as WIN-DAISY, WIN-FIT, NMRSIM, WIN-MAS, WIN-DR and others, which interface with either 1D WIN-NMR or 2D WIN-NMR. This option is only available with the full version of 1D or 2D WIN-NMR.

Process Contains commands and various sub-menus for processing NMR data in different states, i.e. in the time-domain (FID) and in the frequency-domain (spectra). Different commands and submenus are available depending on the type of level and the processing stage.

Analysis This pull-down menu is only available for frequency domain data (spectra) and allows a few simple "analytical" tasks to be performed such as peak picking, calibration, integration or simple spectral analysis.

Output Contains a series of options to prepare your data file for subsequent plotting, including the printer set-up, the inspection and editing of acquisition and processing parameters. It also contains copy and plot/print commands.

Display Offers a set of special options for arranging displays and layouts according to your needs. Basic display functions are also available using the buttons on the button panel, as described below.

Window This pull-down menu is only available with 1D WIN-NMR and is used to select and arrange the four different windows: *1D WINNMR [Spectrum]*, *-[Relaxation]*, *-[Preview]* and *-[Text]* .

Help Call **Help** information about most of the processing options as described in section 4.2.

In this chapter the pull-down menus **File**, **Analysis**, **Output**, **Display** and **Window** will be discussed and illustrated. In chapter 5 the options available in the **Process** menu will be explained.

When 1D WIN-NMR or 2D WIN-NMR is first started, the appropriate maximized application or main display window *1D WINNMR [Spectrum]* and *2D WIN-NMR* respectively appears on screen. Whereas 2D WIN-NMR has only one application window, 1D WIN-NMR has three additional application windows. These four application windows (*Spectrum*, *Preview*, *Relaxation* and *Text*) may be displayed altogether (Multi Document Interface, MDI) on the screen by clicking the MDI window button, or may be displayed pairwise according to your needs by clicking one of the pairs offered in the **Window** pull-down menu. The active application window is indicated by the highlighted title bar (Fig. 4.3).

A window is activated by clicking the left mouse pointer in it, the options available in the menu bar and the buttons in the button panel then change to functions appropriate for the selected window.

Check it in 1D WIN-NMR:

Start the 1D WIN-NMR program and set-up the four MDI/application windows display as shown in Fig. 4.3 by clicking the MDI system menu button. Activate the four windows one after the other and inspect the corresponding button panels and menu bars. Iconize two of the windows and rearrange the other two so that each fills half of the display. To do this use the title bar and/or the window frames or use the **Tile** option in the **Window** pull-down menu. Check the functionality of the minimize/maximize button and of the options offered in one of the system menus. Iconize the 1D WIN-NMR program by clicking the corresponding button in the 1D WIN-NMR title bar.

Check it in 2D WIN-NMR:

Start the 2D WIN-NMR program and check the functionality of the minimize/maximize button. Try to arrange the 2D and 1D WIN-NMR application windows so that each fills half the display. This will be helpful later in the processing of 2D data, where the 1D WIN-NMR options should also be accessible. Iconize the 2D WIN-NMR program by clicking the corresponding button in the title bar.

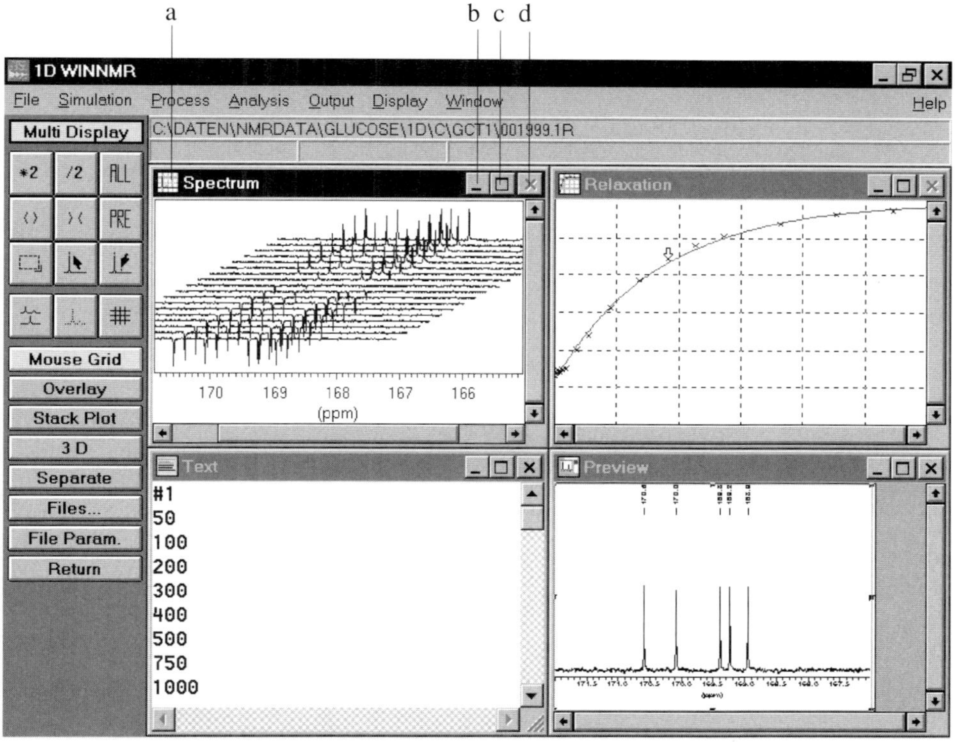

Fig. 4.3: The 1D WIN-NMR application window with all four MDI windows visible. All the MDI windows have: a) a system menu, b) a button to iconize the corresponding window c) a minimize/maximize button and d) a close button. The *Spectrum* window is activated and the corresponding buttons in the button panel are shown.

4.4 File Handling

Whenever 1D WIN-NMR or 2D WIN-NMR are running, clicking the left mouse button when the mouse pointer is over **File** in the main menu bar, opens a pull-down menu relating to various file operations. Fig. 4.4 shows the **File** pull-down menus when activated in 1D WIN-NMR or 2D WIN-NMR. Options to load (**Open...**, **Recall last**) and save (**Save**, **Save as...**) NMR data (FIDs, spectra), to call the WINDOWS file manager (**File Manager**), to start the GETFILE program (**File Transfer**) to copy data files stored on remote computers are available with 1D and 2D WIN-NMR. With 1D WIN-NMR additional options to open job files (**Open Job...**) and to enter a dialog either for serial processing (**Serial Processing**), discussed in chapter 5, or for direct copying and converting files (**Filecopy & Convert**) are available. Finally, in 1D WIN-NMR names of the last data sets loaded are visible for quick access. For 2D WIN-NMR additional options for transmitting single FIDs (**FID-Transmission**), cut out from 2D data matrices, to 1D WIN-NMR exist.

Fig. 4.4: Options of the 1D WIN-NMR (left) and the 2D WIN-NMR (right) **File** pull-down menu.

Check it in 1D WIN-NMR:

Start 1D WIN-NMR and from the **File** pull-down menu choose the **Open...** submenu. Select and load the 1D proton spectrum of glucose D:\NMRDATA\GLUCOSE\1D\H\GH\001999.1R by clicking the **OK** button (Fig. 4.5).

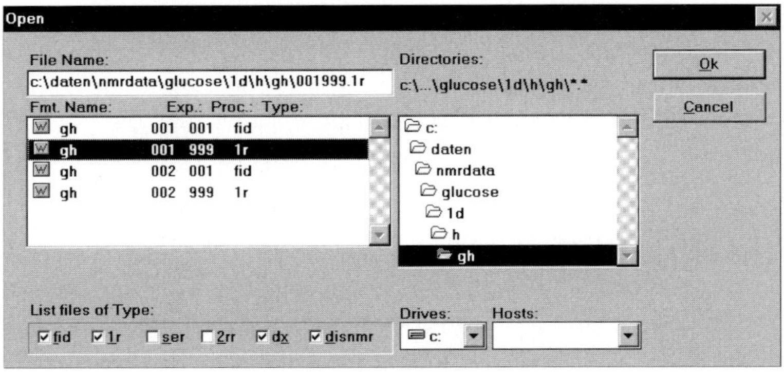

Fig. 4.5: The **Open** dialog box of 1D WIN-NMR.

Using the **Save as ...** option, store the data under the file name D:\NMRDATA\1D\H\GH\001001.1R. Check the various formats listed in the corresponding dialog box, select the binary format and hit the OK button to save the file. Use the **Help** information to familiarize with the **Recall last** function.

Now choose the **File Manager** option in the 1D WIN-NMR **File** pull-down menu which will show the 1D WIN-NMR application window and the WINDOWS file manager window in a dual display mode (Fig. 4.6). In the directory D:\NMRDATA\GLUCOSE\1D\H\GH check that there is a new entry for the 1D proton spectrum stored above.

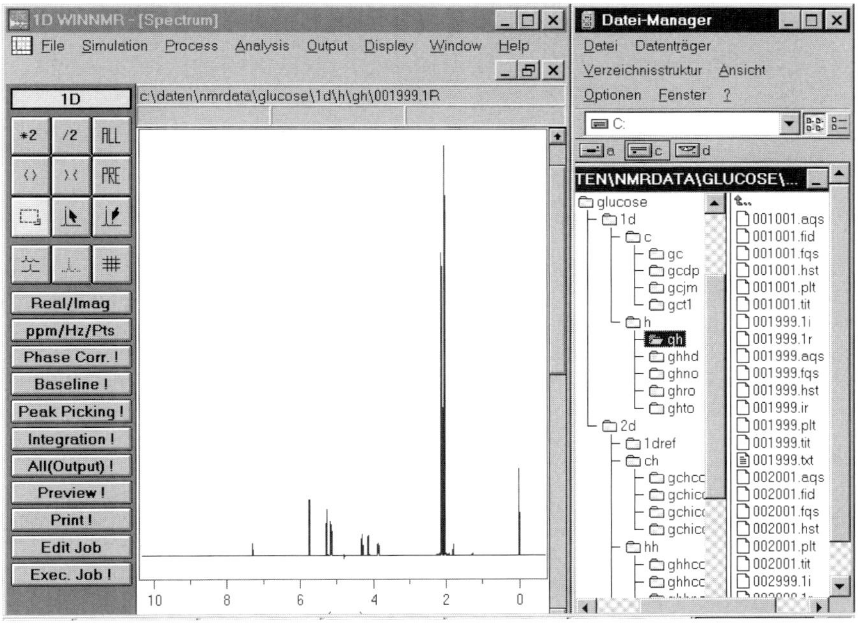

Fig. 4.6: Display as obtained after calling the WINDOWS file manager from within 1D WIN-NMR.

With this dual display still on the screen select in the file manager window the 1D carbon spectrum D:\NMRDATA\GLUCOSE\1D\C\GC\001999.1R and use the drag and drop method to move it directly and most conveniently into the 1D WIN-NMR application window.

Check it in 1D WIN-NMR:

Check reloading of previously displayed data sets. In the box just above the **Exit** command in the 1D WIN-NMR **File** pull-down menu, click on the proton spectrum of glucose D:\NMRDATA\GLUCOSE\1D\H\GH\001999.1R. It will be immediately reloaded and displayed.

Activate **File Transfer** and **Filecopy & Convert** and inspect their dialog boxes. These two options have been discussed in detail in chapter 2.If necessary, use the **Help** routine to get more information about using these options.

Check it in 2D WIN-NMR:

Start 2D WIN-NMR and from the **File** pull-down menu choose the **Open...** submenu. Select the 2D $^1H/^1H$ COSY spectrum of glucose D:\NMRDATA\GLUCOSE\2D\HH\GHHCO\001999.RR) and load it by clicking on this file and the **OK** button respectively. Use the **Save as...** option in the **File** pull-down menu to store the data under the file name D:\NMRDATA\GLUCOSE\2D\HH\GHHCO\001001.RR).

Select the **File Manager** option and check the entries of the new 2D data set.

Check it in WINDOWS:

Use the WINDOWS file manager to load the 2D $^1H/^{13}C$ COSY spectrum D:\NMRDATA\GLUCOSE\2D\CH\GCHCO\001999.RR) using the drag and drop method.

Delete the newly stored 1D and 2D data sets from within the WINDOWS file manager.

4.5 Display of 1D Spectra with 1D WIN-NMR

4.5.1 Buttons with 1D WIN-NMR [Spectrum]

This section describes the function of the most important buttons in the button panel (Fig. 4.2). Some of the additional options available within the **Display** pull-down menu in the main menu bar of 1D WIN-NMR and the function of scroll bars and function keys will be discussed in sections 4.5.2 and 4.5.3 respectively. All affect the display of a spectrum or FID in the *1D WIN-NMR [Spectrum]* window. The display buttons (Fig. 4.7) are available, whenever the 1D WIN-NMR program has been started.
With 1D WIN-NMR the panel title indicates the current operational mode of the button panel, either *1D* for the spectrum window or *Text*, *Relaxation* or *Preview* for the corresponding windows. The function of the buttons in the panel change to reflect the functionality of the selected mode.

Although the symbols on most of the buttons shown below are self-explanatory a brief description of each button follows. For further information you are referred to the **Help** tool.

Increase y-scaling by a factor of 2.

Decrease y-scaling by a factor of 2.

Display complete spectrum.

Expand in the x-direction by a factor of 2.

Contract in the x-direction by a factor of 2.

Toggle between the current display and the most recent expanded part of the spectrum.

The next three buttons change the current mode of the mouse pointer, which affects how the pointer behaves in the window which shows the spectrum. The current mouse pointer mode can be changed by clicking with the left mouse button on one of these buttons (which become highlighted). The current mouse pointer mode is also indicated by the shape of the pointer in the spectrum window:

Button for zooming part of the spectrum. This is the most common mode for the mouse pointer.

Button for placing a spectrum cursor on the top of peaks in the spectrum (*Maximum Cursor* mode). Selecting this mode automatically inserts an info line at the top of the spectrum window which lists information about the pointer position, i.e. the x- and y- co-ordinates of the spectrum cursor in Hz/ppm, the number of the associated data point and its intensity.

Button for placing the pointer at any point in the spectrum (*Perpendicular Cursor* mode). This mouse pointer operates in a similar way to the previously described mode. However the position of the spectrum cursor is determined by a perpendicular drop from the current mouse pointer position onto the spectrum. The spectrum can also be expanded in the x-direction using this mode.

The next buttons in the button panel are highlighted when active and are toggles which may be used to alter how data is displayed in the spectrum window:

This button initiates a *Split-Screen* mode in the window. In the upper trace the full spectrum is always visible and a black box indicates that part of the spectrum, which has been expanded and is visible in the lower trace.

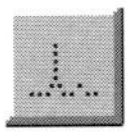

This button displays the data trace as unconnected data points.

This button superimposes a grid on the spectrum display area.

The functions of the additional buttons available with the *1D WIN-NMR [Spectrum]* window are as follows:

Real/Imag Toggles between display of the real and imaginary parts of the spectrum or FID.

ppm/Hz/Pts Toggles between different units for the x-axis.

When an FID or spectrum is read into the spectrum window, the following short-cut buttons for rapid access to the most frequently applied processing, display, edit and plot tasks are available. For most of these buttons a dedicated dialog box to set up and adjust the corresponding parameters is available by clicking with the right mouse button.

Phase Corr.! Carries out a phase correction on your spectrum using the 0th and the 1st order parameters last defined during interactive phase correction (see chapter 5).

Baseline! Carries out a fully automatic baseline correction using the degree of the polynom for the baseline function last defined during interactive baseline correction (see chapter 5).

Peak Picking! Starts peak picking using the parameter PC last set during interactive peak picking (see section 4.6.2).

Integration! Carries out an integration of spectral regions using the same parameters as previously used for automatic integration (see section 4.6.3).

All (Output)! When a region of a spectrum is selected for plotting, this button can be used to scale the peak sizes in the y-dimension so that the largest peak in the spectrum would be on-scale, even if it is not in the viewed region. This enables consistently sized plots to be obtained and is useful when a number of expanded regions of a spectrum should have the same vertical scaling.

Preview! Starts a preview output using previously selected plot options (see section 4.7.3)

Print! Starts hardcopy output using previously selected plot options.

Edit Job! Brings you into a dialogue for editing the “Common Job” file which has been selected via the **Open Job** command from the **File** pull-down menu. For handling this dialogue and using this option you are referred to chapter 5.

Exec. Job! All commands that have been defined in the “Common Job” file will be executed for the current NMR data set. For handling this dialogue and using this option you are referred to chapter 5.

If an FID is read into the spectrum window, special FID short-cut buttons are available. Their functionality will be discussed in chapter 5.

Return This button is not present in the display ground state of the spectrum window, but appears at the bottom of the button panel whenever any other button panel mode is activated via a pull-down menu.
Click this button to exit from the current button panel mode, and return to the display ground state of the spectrum window from which another function with its own button panel mode may be chosen. Selecting an entry from the pull-down menu bar in the main menu automatically causes a change in the display ground state of the spectrum window. The **Return** key on the keyboard has the same function.
The **Return** button behaves the same with the other windows (*Preview*, *Relaxation*, *Text*).

Check it in 1D WIN-NMR:

Start the 1D WIN-NMR program and load the 1D ^{1}H spectrum of glucose D:\NMRDATA\GLUCOSE\1D\H\GH\001999). Try out the various buttons in the button panel to increase/decrease the spectrum intensity, to expand and shrink the spectrum, to display the full spectrum (**ALL** button) or to show the previous expansion (**PRE**).

Check it in 1D WIN-NMR:

Now try to zoom the spectrum by using the zoom button. Test the two cursor modes for setting the cursor on the top of a signal and for positioning it anywhere in the spectrum.

Try out the remaining buttons in this section of the button panel. With the **Split-Screen** button expand part of the spectrum as above and inspect what happens in the upper part of the display.

Check it in 1D WIN-NMR:

Use the same spectrum as above and try out the large buttons in the button panel. First try the **Real/Imag** button. Expand the region of the glucose ring protons (6 –3.5 ppm) and increase the vertical expansion. Try out the other buttons (**ppm/Hz/Pts**, **Peak Picking!**, **Integration!**). Inspect but do not alter the entries in the corresponding dialog boxes, accessible by clicking the right mouse button. Finally click on the **All (Output)!** button which resizes the peaks in your expansion. Increase the vertical scale again and inspect the layout with the **Preview!** button. Print the spectrum displayed using the **Print!** command.

Check it in 1D WIN-NMR:

Activate the **Phase** dialog box and button panel by clicking on the **Phase Correction** submenu in the **Process** pull-down menu. This will change the display ground state and will assign other functionalities to the buttons in the button panel including the **Return** button. Hit this button to return to the **1D** display ground state of the *Spectrum* window. Phasing a spectrum using this option will be discussed in chapter 5.

4.5.2 Additional Display Options with 1D WIN-NMR

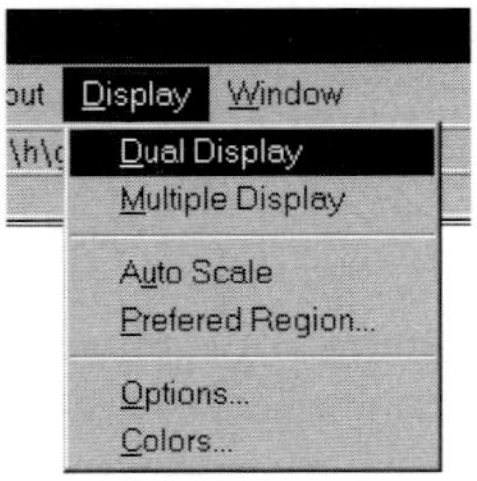

Fig. 4.7: The **Display** pull-down menu of 1D WIN-NMR.

When activating the **Display** pull-down menu (Fig. 4.7) the following functions are available:

Dual Display/ Multiple Display	These commands are self-explanatory and will be used later.
Auto Scale	This command performs the same operation as the **ALL** button in the *Spectrum* window button panel. It resets the x- and y-axis scales so that the complete spectrum/FID is visible.
Prefered Region...	Opens an additional dialog box to define a prefered and frequently used region. This region may be displayed by pressing the panel button **ALL** with the right mouse button.
Options...	Clicking this button opens a new dialog box showing all the various options available for the current display mode of the *Spectrum* window. The contents of this dialog box depend on which display mode is currently active; different contents are visible for the normal, dual and multiple display modes. Use the **Help** tool for more detailed information.
Colors...	This option lets you define colors for the elements in the *Spectrum* window. Again the self-explanatory dialog box opened by this option depends on the currently selected display mode (normal, dual, multiple).

Check it in 1D WIN-NMR:

Load the reference spectrum D:\NMRDATA\GLUCOSE\1D\H\GHNO\1D\008999.1R from the NOE data for glucose. From the **Display** pull-down menu choose the **Dual Display** option and select the NOE-difference spectrum 007999.1R in the same directory.

Try out the functions of the various buttons in the button panel: **Move Trace**, **Reset**, **Show Diff.**, **Separate**, **Options...**, **Return**. In **Options...** set the **Second Trace Factor** to –1. Click on the **Show Diff.** button and estimate the degree of saturation for the irradiated resonance (3.86 ppm). Click on the **Show Diff.** button again to deactivate it. Click on the **Reset** button and select under **Options...**, **Second Trace File** another spectrum, e.g. 006999.1R, in the same directory.

Check it in 1D WIN-NMR:

Again load the reference spectrum (NMRDATA\GLUCOSE\1D\H\GHNO\1D\008999.1R from the NOE data for glucose. From the **Display** pull-down menu choose the **Multiple Display** option and select the whole series of NOE-difference spectra 001999.1r to 008999.1r to inspect the NOEs.

Try out the effect of the buttons in the button panel. The three display mode buttons **Stack Plot**, **3D**, **Overlay**, which toggle between different display modes and the buttons **Mouse Grid** and **Separate**. Use **Files** to selectively delete the files 003999.1R to 005999.1R. Click first on the **File Param.** and then on the **Next Trace** button until the file 009999.1R is selected. Change the **Y-Scaling** to 0.5 and inspect the multiple display again.

Check it in 1D WIN-NMR:

Using the last data set inspect the different **Options** menus available via the **Display** pull-down menu for the normal, the dual and the multiple display modes. Try out the various affects of the various options.

Check it in 1D WIN-NMR:

Load the ^{1}H spectrum of glucose D:\NMRDATA\GLUCOSE\1D\H\GH\002999.1R. Use the **Peak Picking!** button in the button panel for peak picking. From the **Display** pull-down menu choose the **Options ...** submenu. In the dialog box that appears on the screen make sure that the **Peak Labels** box is activated. From the **Display** menu choose now the **Colors...** option and select different colors for the spectrum, the frame, the numbers and the peak labels. Peak picking will be explained in more detail in section 4.6.2.

Check it in 1D WIN-NMR:

Load the ^{13}C spectrum of glucose D:\NMRDATA\GLUCOSE\1D\C\GC\002999.1R. In the **Display** pull-down menu select the **Prefered Regions...** option and set in the corresponding dialog box the **Left Limit** and the **Right Limit** to 100 ppm and 60 ppm respectively. Test the **Prefered Regions...** option by clicking the **ALL** panel button alternately with the left and right mouse button.

4.5.3 The Use of Scroll Bars, Keys and Function Keys with 1D WIN-NMR

The scroll bars in the *Spectrum* window let you scroll the spectrum in the x- and y-directions. The length of the horizontal and vertical bars indicates how much of the complete spectrum is displayed. The scroll bars are used in the standard way as for any MS-WINDOW program.

The + and – keys on the keyboard increase and decrease the y-scale of the spectrum by a factor of 2 and correspond to the ***2** and **/2** buttons respectively.

The function key **F1** calls the **Help** function. **F5** expands and **F6** compresses the spectrum in y-direction. The spectrum can be shifted up with **F7** and down with **F8,** to the left with **F9** and the right with **F10**.

Check it in 1D WIN-NMR:

Use the last data set to try out and test the use of scroll bars, **+** and **–** keys and the function keys. Use the **Help** information if necessary.

4.6 Basic Processing Steps with 1D Spectra

Assuming your 1D NMR data has already been processed, i.e. the Fourier transformation has been performed, the phases of the signals have been correctly adjusted to pure absorption and your 1D spectrum is stored on your PC's disk, there are still a few final processing steps to be performed, before the final layout is completed and the data can be plotted.

4.6.1 Calibration

Spectrum calibration, for which a reference compound such as tetramethylsilane (TMS) or – less preferably – the solvent signal is used, allows the chemical shifts of the investigated compound to be compared with those of reference compounds, available in the literature or in commercial data bases.

Calibration is accessible via the **Calibration** option in the **Analysis** pull-down menu (Fig. 4.8). Its activation switches the button panel into the *Calibration* mode, which allows you to select between two different calibration methods.

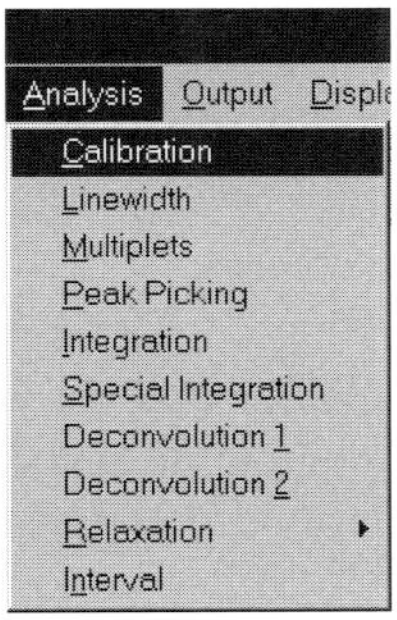

Fig.4.8: **Analysis** pull-down menu

Before activating the **Calibrate** button, the cursor should first be fixed on the desired position in the spectrum, using either the *Perpendicular Spectrum* or more preferably the *Maximum Spectrum* or cursor mode. Clicking the **Calibrate** button opens a dialog box which allows the **X:** value in ppm and the **Y:** value in cm of the cursor position to be entered provided that **File Compare Mode** for both x and y are disabled (Fig. 4.9).

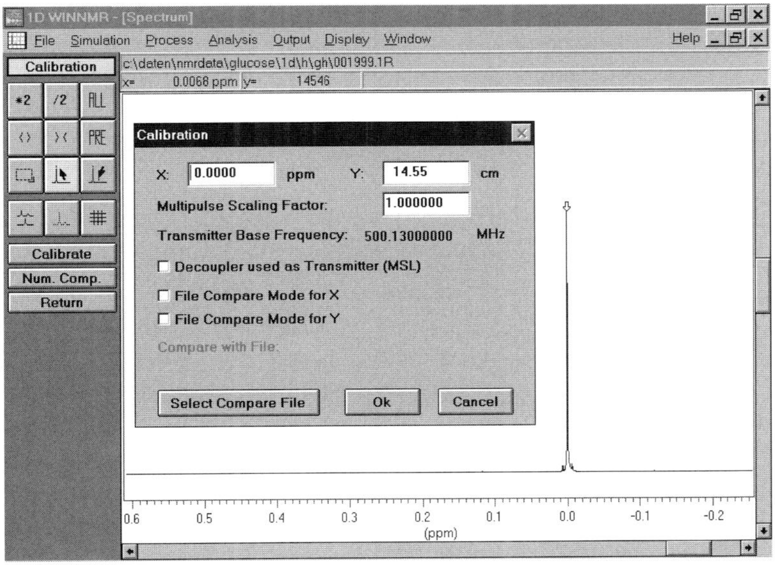

Fig. 4.9: The **Calibration** dialog box.

It is also possible to calibrate the spectrum using a reference spectrum stored on disk by using the **Select Compare File** button within the **Calibration** dialog box. This is particularly useful if the spectrum being displayed is an expansion and does not contain the reference signal (e.g. TMS) and you wish to compare it with a separately measured spectrum showing the whole spectral region.

For further details about the **Compare File** for x or y feature and for more information regarding the items accessible with the **Num. Comp.** button or for performing an offset-independent calibration of a spectrum via the **Edit Job** button, you are referred to the corresponding **Help** information.

Check it in 1D WIN-NMR:

Load the 1D ^{1}H spectrum of glucose D:\NMRDATA\GLUCOSE\1D\H\GH\001999.1R and calibrate the spectrum with respect to the TMS signal using the **Calibration** option in the **Analysis** pull-down menu. Expand the TMS signal, set the cursor on the top of the peak and click on the **Calibrate** button to calibrate this peak to 0 ppm. Keep in mind the new shift value for the anomeric proton (5.724 ppm). Store the spectrum under its original name.

Now load the expanded ^{1}H spectrum of glucose D:\NMRDATA\GLUCOSE\1D\H\GH\002999.1R and calibrate this spectrum with respect to the spectrum stored above. Use the **Select Compare File** button in the **Calibration** dialog box and use the calibrated spectrum D:\NMRDATA\GLUCOSE\1D\H\GH\001999.1R as the reference file. Activate the **File Compare Mode for x** box. Check the shift value of the anomeric proton before and after the calibration process initialized by clicking the **OK** button. Store the calibrated spectrum under its original name.

Both 1D ^{1}H spectra of glucose are now stored as calibrated files. Use the dual display to investigate both spectra and to check the identical calibration.

Check it in 1D WIN-NMR:

Calibrate all the other ^{1}H 1D glucose spectra stored in the directory D:\NMRDATA\1D\H, i.e. GHDC, GHNO\1D, GHRO and GHTO. Use the **Select Compare File** button as described above. Do not forget to store the spectra after calibration under their original names.

Check it in 1D WIN-NMR:

Load the 1D ^{13}C spectrum of glucose D:\NMRDATA\GLUCOSE\1D\H\GC\001999.1R, check and eventually calibrate the spectrum with respect to the TMS signal. Proceed as above and calibrate this peak to 0 ppm. Store the spectrum under its original name.

Load the expanded ^{13}C spectrum of GLUCOSE D:\NMRDATA\GLUCOSE\1D\H\GC\002999.1R and calibrate this spectrum with respect to the spectrum stored above. Use again the **Select Compare File** button in the **Calibration** dialog box and use the spectrum D:\NMRDATA\GLUCOSE\1D\H\GC\

001999.1R as the reference file for this purpose. Store the calibrated spectrum under its original name.

Both 1D ^{13}C spectra of glucose are now stored as calibrated files.

Check it in 1D WIN-NMR:

Calibrate all the other ^{13}C 1D glucose spectra stored in the directory D:\NMRDATA\1D\C, i.e. GC, GCDP, GCJM and GCT1\1D. Use the **Select Compare File** button as described above. Do not forget to store the spectra after calibration under their original names.

4.6.2 Peak Picking

Peak picking is usually performed to measure chemical shifts (in ppm relative to a reference peak in the spectrum) or to measure line separations (in Hz) in multiplets in order to calculate or at least to estimate coupling constants.

For rapid peak picking the button **Peak Picking!** in the button panel may be used. A corresponding dialog box, similar to the dialog box shown in Fig. 4.10, for adjusting the peak picking parameters may be opened by clicking the right mouse button, otherwise the parameters of the last interactive peak picking are applied.

The **Peak Picking** option in the **Analysis** pull-down menu allows peak picking in an interactive way. Activating this menu option switches the button panel into the *Peak Picking* mode, the mouse cursor to *Peak Picking Zoom* mode and opens a dialog box (Fig. 4.10).

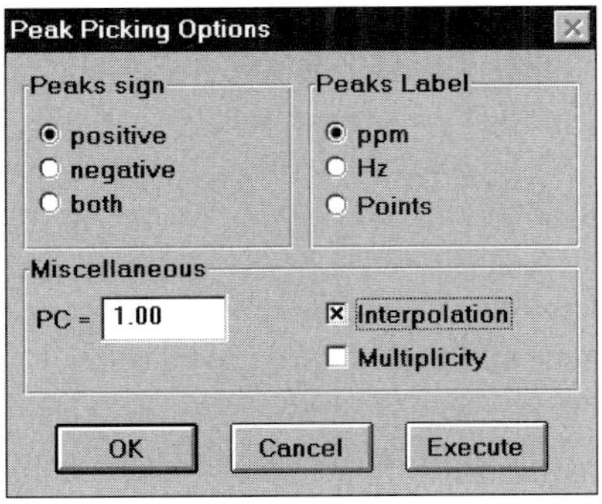

Fig. 4.10: The **Peak Picking Options** dialog box.

In this dialog box the **Peaks sign**, the **Peaks Label** and the parameter **PC**, which determines the sensitivity of the peak picking algorithm may be specified. Furthermore, the *Interpolation* and *Multiplicity* mode can be set which increases the accuracy of the peak picking and lists the multiplicity information respectively.

The buttons in the peak picking button panel have several functions:

Edit Cursor Allows you to edit (delete, add multiplicity information) currently displayed peak labels. During this operation the cursor will only move to marked peaks.

Mouse PP The *Mouse Peak Picking* mode allows you to define the x-and y-limits for the peak picking regions (Fig. 4.11). It is possible to pick peaks in many different regions of the spectrum using this option.

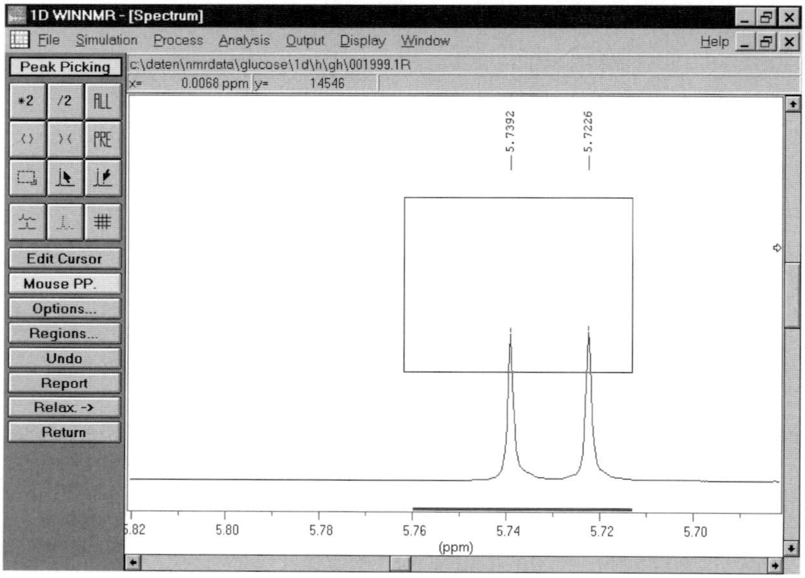

Fig. 4.11: Peak picking in the *Mouse Peak Picking* mode.

Options... A dialog box allows you to inspect and edit the peak picking parameters.

Regions... A dialog box allows you to list, edit, load and save peak picking regions and to execute peak picking according to the entries.

Undo Erases all the individual peak picking regions and all the picked labels. A region is created that spans the complete spectrum.

Report A list box is shown which includes the peak number, the data point number, the frequency [Hz], the chemical shift [ppm], the intensity of the peak, either absolute or relative to the largest peak. Several options for editing and for copying the list into the clipboard, for exporting it into the *Preview* window or into other WIN-NMR applications programs, for saving it, for specifying and setting-up the current printer and for printing it are available.

Relax.-> Allows the *Peak Picking* mode to be carried over into the *Peak Picking - Relaxation* mode. For peak picking in connection with relaxation studies, you are referred to *Modern Spectral Analysis* (volume 3 of this series).

Check it in 1D WIN-NMR:

Load the ^{1}H spectrum of glucose D:\NMRDATA\GLUCOSE\1D\C\GH\001999.1R and perform an interactive peak picking in Hz for the ring protons using the **Peak Picking** option in the **Analysis** pull-down menu.. Do not select the **Interpolation** option in the first dialog box at this stage. Try out the *Mouse Peak Picking* mode (Fig. 4.11). Click on the **Mouse PP** button and selectively peak pick the signals of the glucose ring protons (6 – 3.5 ppm). This button works in exactly the same way as the mouse zoom button. Exit the *Mouse Peak Picking* mode. If necessary expand the region to inspect the result in the spectrum. Check the entries in the peak picking regions list (**Regions...**) and the report peaks list (**Report**) by clicking the appropriate button. Before exiting the *Peak Picking* mode save the selected region for use with later examples. In the **Peak Picking Regions** dialog box select the **Save** option and store this region file with the name D:\NMRDATA\GLUCOSE\1D\C\GH\001999.RGN.

Now select the *Interpolation* mode in the **Peak Picking Options** dialog box and repeat the procedure. Compare the results for the different methods on the screen and in the corresponding peak lists. Expand the spectrum in the region of a peak and inspect the effect of the interpolation.

Check it in 1D WIN-NMR:

Load the normal ^{13}C spectrum of glucose D:\NMRDATA\GLUCOSE\1D\C\GC\001999.1R. Enter the quick mode of peak picking (in ppm) using the **Peak Picking!** button in the button panel. Click first with the right mouse button on it to inspect and eventually adjust the parameters and then with the left mouse button to initialize peak picking. This will label all peaks, including a few noise peaks, of your spectrum with the corresponding ppm values. From the **Analysis** pull-down menu choose the **Peak Picking** option. Close the dialog box that appears. Click on the **Edit Cursor** button and move the cursor onto one of the lines of the $CDCl_3$ triplet.Click the left mouse button to position the cursor and the right button to remove the label. Selectively delete the labels for the remaining $CDCl_3$ signals and for any noise peaks. Save the selected region in a peak picking region file for use with later examples.

Check it in 1D WIN-NMR:

Load the DEPT spectrum of glucose D:\NMRDATA\GLUCOSE\1D\C\GCDP\003999.1R. Use either **the Peak Picking!** button together with the corresponding dialog box for the quick mode, or choose the **Peak Picking** option from the **Analysis** pull-down menu. In the **Peak Picking Options** dialog box set the **Peaks sign** to **both** (negative and positive peaks) and the **Peaks Label** to **ppm**. Click on the **OK** button.

Switch the cursor to the *Maximum Cursor* mode and label a few signals with the cursor "by hand". Click the left mouse button to position the cursor and the right button to label the peak. Try to selectively delete one of the labels again. Inspect the result in the spectrum and in the **Report: Peaks** dialog box opened with the **Report** button. Save the peak picking regions in the corresponding region file.

Load the other ^{13}C DEPT spectra of glucose using the previously stored region file and perform a peak picking for the signals in the same region as above.

4.6.3 Integration

Integration of NMR spectra – almost exclusively applied in ^{1}H NMR – is used to determine the areas of the individual resonances and to calculate relative ratios of nuclei in particular chemical environments. Integration is also used to calculate NOEs or to follow time dependent phenomena, e.g. the progress of a chemical reaction in the NMR tube. The accuracy of integration is affected by a series of factors which should not be underestimated. The reliability of integrals depends upon the proper settings of several acquisition and processing parameters. Some of the most important parameters are the recovery delay during acquisition, the type and parameters of any weighting function and the quality of phasing in the spectrum. The accuracy of integrals is also influenced by the appearance of the spectrum itself. Integration of partially overlapping signals in highly overcrowded spectral regions, of small signals close to very large signals and of sharp signals in the presence of very broad signals is more demanding and sets upper limits for the accuracy.

For integration of 1D spectra several possibilities exist:

- For rapid integration click with the left mouse button on the **Integration!** button in the button panel. The same parameters as set for the automatic integration routine available in the **Analysis** pull-down menu (see below) will be used and individual integral traces for each multiplet will automatically be calculated and shown on the display. If you want to adjust the integration parameters before the integration is initialized, first click with the right mouse button on the **Integration!** button to open a corresponding dialog box.
- A fully-interactive *Integration* mode is available with **Integration** in the **Analysis** pull-down menu. Selecting this **Integration** option will switch the button panel into *Integration* mode and sets the mouse into the *Perpendicular Spectrum Cursor* mode which allows the integration regions to be defined interactively (Fig. 4.12).

The integration buttons panel has several functions:

Slider Var. Allows you to select which integral parameter is assigned to the slider function. The currently displayed parameter can then be adjusted either for the active integral (marked by a solid bar) or for all integrals simultaneously. The prefix **Comm.** or **Indiv.** before the parameter indicates that all integrals or just the selected integral will be adjusted.

Four integral parameters exist **Bias**, **Slope**, **Offset** and **Scale** with the parameter currently being adjusted indicated in the title bar of the slider box. The **Bias** and **Slope** parameters are the 0th and 1st order coefficients of a straight line defining the baseline in the integral region. These two parameters should be adjusted until the integral is flat at both ends. A bias correction is usually necessary if the spectrum baseline deviates from the zero horizontal line by a frequency independent offset. A slope correction is usually necessary if the spectrum baseline is not flat throughout. This mainly occurs with bad phased spectra (seer chapter 5) or may be observed in spectra dominated by a very intense solvent signal. Before applying a bias or slope correction it is recommended, however, to perform a baseline correction and/or to carefully rephase the spectrum. A quick and most convenient way to correct the spectrum baseline prior to integration is offered in the integration **Option** dialog box. The **Offset** parameter moves the integrals in the y-direction and the **Scale** parameter controls the size of the integrals. If an individual integral is scaled differently relative to the rest of integrals, a scaling factor is listed above, which is useful in cases where there are intense and weak signals in your spectrum.

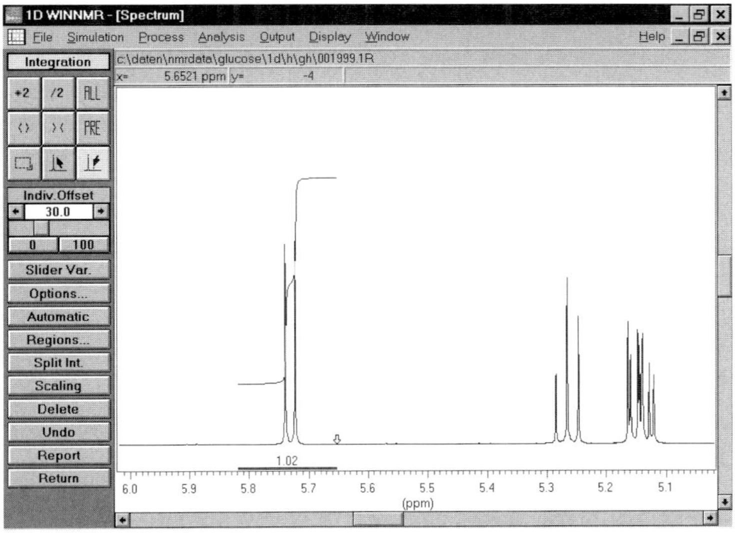

Fig. 4.12: Display and button panel in the *Integration* mode.

Options Opens a dialog box with options for assigning the area value in the edit box either to the sum of all integrals or to the selected integral (**Manual Calibration**), for selecting integral regions (**Selected Integral Region**), for calibrating of integrals with integrals of a stored file (**File Calibration**), for adjusting the common integral offset and the common or individual integral scaling (**Display**) and for enabling an automatic baseline correction prior to integration and controlling the panel slider (**Miscellaneous**).

Automatic Calls a dialog box to control the automatic integration, which can be carried out for either the whole spectrum or for automatically fixed regions. Automatic detection of integration regions is controlled by four parameters. Use the **Help** option for details.

Regions Opens the **Integral Regions** dialog box with a list of all of the defined integrals with their limits and individual display scaling. You may edit individual fields in this list box, save and restore integral ranges to and from files on disk and insert or delete intervals according to your requirements.

Split Int. Toggles the **Split Integral** option on and off and allows you to interactively split integral regions into several smaller regions.

Scaling When clicking this button the height of the active integral is set to the height of the corresponding peak and all the other integrals within the spectral window are scaled.

Delete Deletes the active integral.

Undo Deletes all of the defined integrals.

Report This button opens the **Report Integrals** dialog box where the numerical results of the integration process are summarized. The list consists of the limits for all integrals in ppm and Hz together with the normalized and absolute integral areas. Further manipulation of this list is possible using the buttons on the right side of the dialog box. These buttons have already been explained in the peak picking section 4.6.2.

- A further integral mode is available with **Integration** in the **Process** pull-down menu. This, however, will simply replace the spectrum by its integral. Its main purpose is to inspect the integral of the whole spectrum before actually and interactively execute the integration. Use the **Help** option for more informations if necessary.

Check it in 1D WIN-NMR:

Load the proton spectrum of glucose D:\NMRDATA\GLUCOSE\1D\H\GH\001999.1R and expand the region with the glucose signals.

Use first the left mouse button to click the **Integration!** button in the button panel and perform a quick automatic integration. Click then the **Integration!** button with the right mouse button to try the option for rapid parameter adjustments and start the integration with the **Execute** button in the corresponding dialog box.

In the **Analysis** pull-down menu choose the **Integration** option. Clicking on the **Automatic** button in the corresponding button panel opens a dialog box. Select the *Detect Area* mode and click on the **Execute** button to perform an automatic integration.

Keep in mind the appearance of these integrals or store them by clicking the **Save as...** button in the **Integral Region** dialog box accessible via the **Regions...** button in the button panel. Note that the extension for integral regions files is IR. Cancel the integrals with the **Undo** button.

Check it in 1D WIN-NMR:

From the **Process** pull-down menu choose the **Integration** option. Using the same spectral region as above, try out the **Execute** command and the *Interactive* mode of integration.

Check it in 1D WIN-NMR:

From the **Analysis** pull-down menu choose the **Integration** option. Using the same spectral region as above, perform an integration over the entire region. With the cursor in the *Perpendicular Spectrum Cursor* mode define the left end and the right end with the cursor to give one single integral. (To define a limit and to position the cursor click the left hand mouse button; to show the integral curve click the right mouse button.) In the **Options...** dialog box make sure that the 'Baseline Correction' option is not activated. Click on the **Scaling** button to set the integral to full scale. Adjust the integral until it is flat at both ends using the **Bias** and **Slope** correction options. To adjust the integral baseline click on the **Slider Var.** button until the 'Indiv. Bias' or the 'Indiv. Slope' caption appears and adjust the integral bias and slope respectively using the slider dialog box. If necessary, choose the lowest sensitivity for these two parameters (–0 / 0). The integral should have no slope in the regions of the spectrum that do not contain any signals. If necessary consult the **Help** menus.

Note: Equal slopes of the integral baseline may by corrected with the **Bias** option. It is recommended to apply a spectrum baseline correction (see chapter 5) or to simply activate the **Basline Correction** option in the integral **Options** dialog box prior to integration.

Unequal slopes of the integral baseline may be corrected either by the **Slope** option or a combined application of the **Slope**/ **Bias** options.

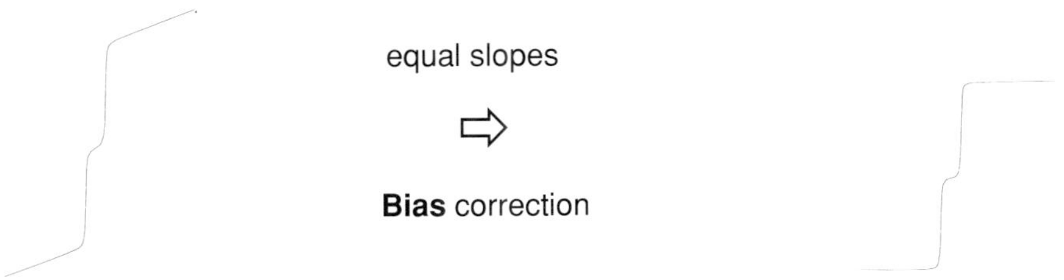

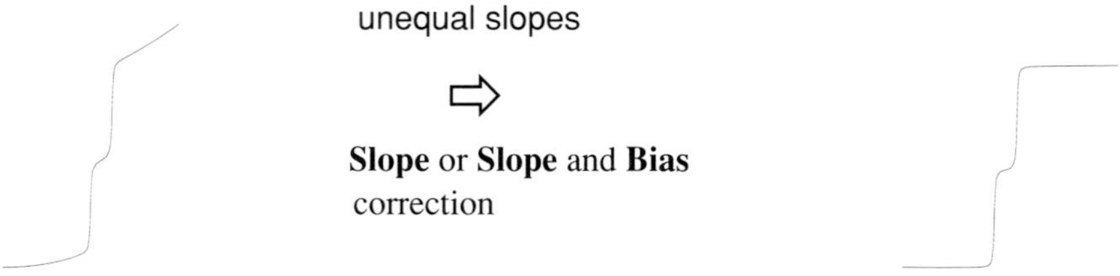

Repeat the integration but activate the *Baseline Correction* mode in the **Options...** dialog box, and compare the two results.

To calibrate this single integral choose the **Options...** dialog box and set the **All Integrals** parameter to the value 22, the total number of protons for peracetylated glucose.

Use the **Split Int.** function to split the single integral into a series of integrals. (Move the cursor to the required position and click the left then the right mouse button.) Select wider regions for each integral compared to the automatic mode and, if possible, reset the integrals after each multiplet. Delete the "zero-integrals" in the signal-free regions using the **Delete** button. In the **Options...** dialog box disable the **Panel Slider Control for all Integrals** option. Select one of the integrals and perform several operations e.g. scaling, offset, bias correction, ... Enable the **Panel Slider Control for all Integrals** option and repeat the same operations.

Activate the **Options...** button again and set the Common Offset to 10%. Try out the other **Display** and **Miscellaneous** options in the dialog box. Select various integrals and try out the **Scaling** button. Adjust the integrals to suit your requirements.

Click on the **Regions...** button. In the dialog box that appears on the screen click on the **Save...** button. In the new dialog box that appears enter the filename D:\NMRDATA\GLUCOSE\1D\H\GH\001001.IR) and save the integral regions for use in later applications.

Check it in 1D WIN-NMR:

Load the reference file from the 1D NOE experiment D:\NMRDATA\GLUCOSE\1D\H\GHNO\1D\008999.1R. From the **Analysis** pull-down menu choose the **Integration** option. Use the **Regions...** button to load the previously stored integral regions file. If necessary adjust the integral regions. Calibrate the integral corresponding to the seven ring protons of glucose to have a total integral of 7. Store the new integral regions in a regions file. Now load one of the NOE difference spectra. Using the **Regions...** button load the regions file of the NOE reference file. Click on the **Options...** button. In the dialog box that appears on the screen, click on the **Select Calibration File** button and specify the NOE reference file as the calibration file. The integrals are now calibrated with respect to the NOE reference file allowing the calculation of the degree of saturation of the selectively preirradiated multiplet. Together with the other integral values of the reference spectrum and the integrals of the NOE enhanced peaks, the corresponding NOE values may be calculated as well.

Attention!: Before using another spectrum as an integral calibration file, the integral regions for this new reference must first have been defined and stored and the spectrum and corresponding integral plotted!

- A last interactive *Integration* mode is available with **Special Integration** in the **Analysis** pull-down menu. Selecting this integration option will switch the button panel into *Special Int.* mode and a few dedicated panel buttons are available. This mode is used for manually integrating a region of the spectrum that contains overlapping peaks. For further details on using these special features consult the appropriate **Help** routine.

4.6.4 Simple Spectral Analysis

The first step in the analysis of any multiplet structures present in a NMR spectrum is usually a simple first order approach, ignoring any second order effects, to determine the approximate chemical shifts, coupling constants and the coupling patterns. Although in principle this type of analysis is only strictly applicable in the case of simple first order spectra i.e. where the difference in chemical shifts between the coupled nuclei is at least one order of magnitude greater than the corresponding coupling constant, this method of spectral analysis is very popular and widely used.

In those cases where the appearance of the spectrum is more or less first order, but the rules are no longer strictly fulfilled, this kind of spectral analysis can still be useful. Although the true chemical shifts and coupling constants can no longer be directly evaluated from the centres of multiplets and from line splittings respectively, this simple type of analysis helps to "understand" the structure of the coupling trees. In conjunction with linewidth information (see the end of this section) the results of such a simple evaluation may be used as input values for a more rigorous quantum mechanical analysis using the WIN-DAISY software tool. The analysis of NMR spectra using WIN-DAISY is described in detail in *Modern Spectral Analysis*, volume 3 of this series.

The **Multiplets** option in the **Analysis** pull-down menu provides a number of graphical tools to allow you to perform a complete first order analysis of spectra that contain up to nine different coupling levels. Using this function NMR spectra displaying both homonuclear and heteronuclear coupling systems may be analysed. The result of such an analysis is displayed as a coupling tree and may either be plotted together with the corresponding spectrum or printed in table form. The **Multiplets** function has options that allow either manual or automatic assignment of coupling partners. The results of such a multiplet analysis can be stored together with the spectrum as a MLT file or be directly exported or linked with other WIN-NMR application programs e.g. WIN-

DAISY, STRUKED. Storing the multiplet file allows the results to be inspected or an incomplete analysis to be finished at a later stage.
Activating the **Multiplets** option switches the button panel into the *Multiplet* mode (Fig. 4.13) containing a number of different options. Most of these options plus some additional ones are also accessible in the pop-up menu and submenu's that appear when the mouse cursor is positioned inside the display window and the right mouse button clicked.

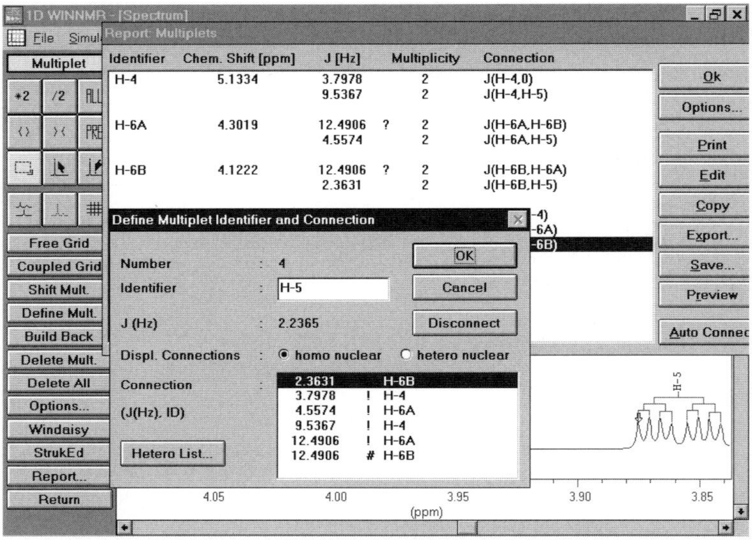

Fig. 4.13: Display and button panel in the *Multiplet* mode with the **Report: Multiplets** dialog box and the box for manual connection and annotation activated.

The buttons in the **Multiplet** button panel have several functions:

Free Grid With the spectrum cursor in the *Maximum Cursor* mode and set on the apex of a selected peak this option allows a predefined multiplet grid to be graphically adjusted to the multiplet lines of the spectrum. Before using this option, the multiplicity of the grid (**2 Distance Line**, **3 Distance Line**, ..., **9 Distance Line**) must be defined. This may be achieved either by clicking the right mouse button in the display window and selecting the *Free Grid* mode in the pop-up menu or by clicking on the **Options...** button in the button panel and adjusting the **Distance Lines** parameter in the **Multiplet Options** dialog box.

Define Mult. The multiplet grid set up with the **Free Grid** button will be defined and assigned to the selected lines in the spectrum. The grid appears in red on the screen indicating that it is the active grid.

Shift Mult. This allows the active multiplet or multiplet tree grid to be shifted vertically.

Coupled Grid With the appropriate multiplet grid activated (with the spectrum cursor in either the *Maximum-* or the *Perpendicular Cursor* mode, position the cursor on the required multiplet and click the left mouse button) a duplicate grid is generated that may be dragged (holding down the left mouse button) to the required position in the spectrum. Releasing the left mouse button freezes the duplicated multiplet grid at its current position where it will be connected with the originally activated multiplet grid to form a coupling tree.

Build Back Allows the coupling tree to be build back in a step-wise fashion to the original multiplet grid (lowest level in the tree hierarchy).

Delete Mult. Deletes the activated coupling tree or multiplet grid.

Delete All Deletes all coupling trees and/or multiplet grids.

Options... This opens the **Multiplet Options** dialog box where the number of **Distance Lines**, the **Capture Range**, **Drift Range**, **Min. Intensity**, **Min. Delta/J** (first order criteria) and the orientation of the multiplet labels may be selected.

Report... This button opens the **Report: Multiplets** dialog box containing the coupling tree structure together with the corresponding "chemical shifts" (multiplet centres) in ppm, "coupling constants" (multiplet splittings) in Hz, the multiplicities and the connections in table form. Buttons on the right hand side of this dialog box allow this report to be printed, edited, exported and saved. Coupling connections may be established either automatically by clicking the **Auto Connect** button or manually; selecting a line in the report and double clicking with the left mouse button opens a further dialog box where the appropriate connection may be selected.

StrukEd Allows you to connect the multiplets with the atoms in the molecular structure using the STRUKED program.

Windaisy Allows you to export the evaluated chemical shifts and coupling constants as starting values to the WIN-DAISY program for a more rigorous quantum mechanical spectral analysis.

For further information use the **Help** routine.

Check it in 1D WIN-NMR:

Load the proton spectrum of peracetylated glucose D:\NMRDATA\GLUCOSE\1D\H\GH\002999.1R. Expand the region 4.4 – 3.7 ppm which contains the signals of the ring protons H-6A, H-6B and H-5. From the **Analysis** pull-down menu choose the **Multiplets** option and generate a coupling tree for all three multiplets. Expand a multiplet until all the lines are clearly visible; with the cursor in *Maximum Cursor* mode set the spectrum cursor on one of the multiplet lines and use the **Free Grid** and **Define Mult.** buttons option to set up the first multiplet grid. Use the **Coupled Grid** button to duplicate the original multiplet grid and continue to build up the coupling tree step by step. Repeat this procedure for the other multiplets. Annotate the three multiplets and explore the options available from within the **Report: Multiplets** dialog box. Compare your result with the result stored with the already analysed spectrum D:\NMRDATA\GLUCOSE\1D\H\GH\002998.1R.

The **Linewidth** option in the **Analysis** pull-down menu allows you to measure the linewidths at half height. The ability to recognize different linewidths in your spectrum is important because it may indicate additional molecular processes going on in solution. Broadening of some of the resonances may be indicative of additional non-resolved couplings, dynamic processes or different types of relaxation mechanisms selectively affecting a particular observed nucleus. This option will also be used to estimate linewidths for use as input data for WIN-DAISY as described in detail in *Modern Spectral Analysis*, volume 3 of this series.

Check it in 1D WIN-NMR:

Load the proton spectrum of peracetylated glucose D:\NMRDATA\GLUCOSE\1D\H\GH\002999.1R and expand a methyl resonance. From the **Analysis** pull-down menu choose the **Linewidth** option. With the cursor in *Maximum Cursor* mode, select a resonance line by clicking the left mouse button. With the cursor set on the top of the peak, click the right mouse button. A horizontal line indicates the intensity at half height and the corresponding line width (in Hz) will appear below the title bar.

4.7 Plotting 1D Spectra

Plotting NMR spectra is never a job of secondary importance. To obtain the best results a few rules should be followed when preparing the page layout. Besides the spectrum itself, the layout should contain all relevant information, including the acquisition and processing parameters and a spectrum title. The acquisition and processing parameters are necessary to interpret and "understand" the spectrum and to allow the experiment to be repeated under the same conditions to reproduce and confirm the results. These parameters are also needed when preparing the experimental part of your thesis or a publication. The spectrum title should immediately identify the sample that has been measured and the experiment that has been performed.

The quality of plots strongly influences the interpretation and spectral analysis and the layout should therefore be adjusted to these subsequent tasks. In most cases an overview plot i.e. 0 to 10 ppm for a proton spectrum, will not suffice and suitable expansions e.g. 5Hz/cm or 10Hz/cm should be plotted as well. Signals should be labelled in either ppm or Hz depending on the problem. Where applicable integrals should be plotted and calibrated. If a series of spectra measured with the same experiment have to be plotted, standard layout(s), designed to your requirements, should be prepared and used throughout. This not only speeds up your daily plotting tasks, but simplifies comparative spectral interpretations. 1D WIN-NMR as well as 2D WIN-NMR offers a variety of features for this purpose.Last but not least, the plot layout, especially if used for presentations, should also be satisfying from an aesthetic point of view.

For plotting 1D spectra two modes exist:

- For rapid plotting click with your left mouse button on the **Print!** button in the button panel and the same layout and plotting parameters as used for the last plot will be applied. To most conveniently adjust the layout and plotting parameters in this rapid plotting mode, click first with your right mouse button on the **Print!** button, which will open a corresponding dialog box.
- For plotting in a more interactive mode the corresponding options in the **Output** pull-down menu will be used.

This section gives an overview of all the options for plotting spectra available in the **Output** pull-down menu (Fig. 4.14) of 1D WIN-NMR. These options affect the output of spectra and related lists (parameter lists, pulse-program lists, or spectral processing history lists). Spectra and lists are usually first displayed and inspected on the screen in the *1D WIN-NMR [Spectrum]* window prior to being sent to a hardcopy device. The *Preview* window shows an exact copy on the screen of your plot (WYSIWYG-philosophy). It is highly recommended to exploit this important feature before plotting your NMR spectra in order to avoid wasting paper and time.

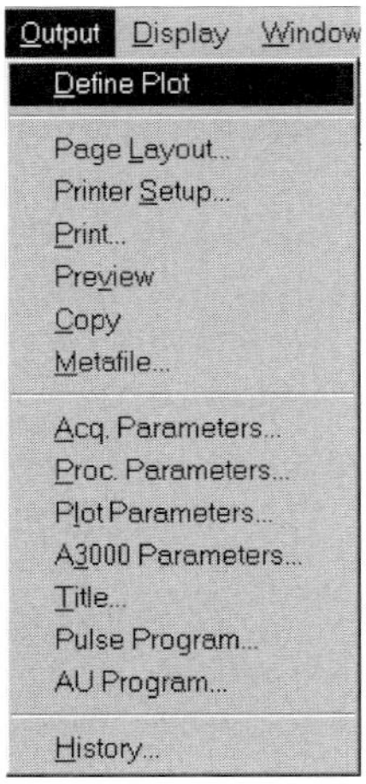

Fig. 4.14: **Output** pull-down menu.

In order to prepare a plot layout the NMR data must first be loaded, the spectrum must be displayed on the screen and any final processing e.g. spectrum calibration, must be performed. The arrangement of the spectrum display on the screen is also the first step in preparing the page layout for the final plot. Thus most of the display options discussed in section 4.6 will also be used in this section. This section will also discuss the additional features relating to setting up and modifying the page layout as well as the various plotting options available with 1D WIN-NMR.

4.7.1 Define Plot

Activating the **Define Plot** command in the **Output** pull-down menu sets the button panel in the spectrum window to *Plot* mode. The functions of the various buttons in the button panel and the effect of the entries in the dialog boxes they open are as follows:

Calibrate... Opens the **Calibration** dialog box (Fig. 4.9) discussed in section 4.6.2.

x-Area... Clicking this button opens the **Plot, x-Area** dialog box (Fig. 4.15), which allows you to specify the frequency limits for the plot and the width of the plot in cm. Activating the *Auto Arrange to Paper* or *the use current Preview Layout* mode disables the **Picture Width** field. In this case the plot width is set automatically according to the selected plot format or the active layout in the *Preview* window (see below) respectively. For the latter option, the layout must have already been loaded into the *Preview* window, otherwise this option is disabled. Using **File Compare Mode** it is possible to output a spectrum with the same x- and y-scales as used in previous plots making it easier to compare spectra. Before **File Compare Mode** is activated, click on the **Select Compare File** button to select a reference spectrum from disk. When **File Compare Mode** is activated, it is not possible to edit any of the fields in this dialog box.

Note: For **File Compare Mode** to work properly, the reference spectrum must be saved to disk immediately after it has been plotted.

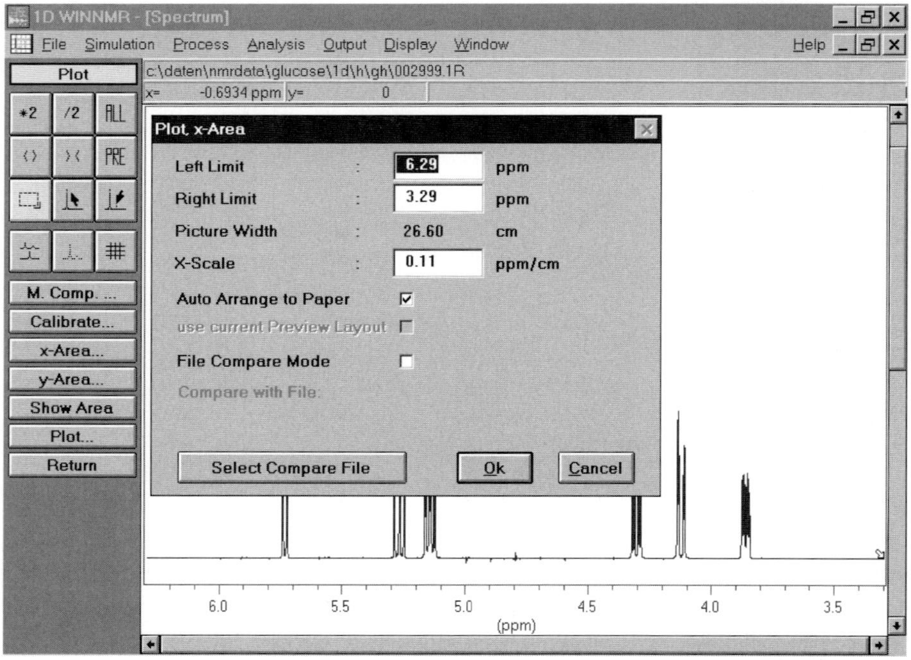

Fig. 4.15: The **Plot, x-Area** dialog box.

y-Area... Selecting this button opens the **Plot, y-Area** dialog box, which allows you to set values for the intensity limits of the plot and the **Picture Height** of the plot in cm. Again **File Compare Mode** allows the y-parameters to be adjusted to those of a reference spectrum stored on disk (see **x-Area**).
Use the **Help** tool for more information regarding the parameters, **Offset** and **Y-Scale**.

Show Area After defining a plot area by means of the **x-Area...** and **y-Area...** buttons, the user can still change the display of the spectrum in the *Spectrum* window. The **Show Area** button resets the display in the *Spectrum* window to the same as in the plot.

Plot... Opens the **Page Layout** dialog box (Fig. 4.16) discussed in section 4.7.2.

Use the **Help** tool for more information regarding the function of the **M. Comp. ...** button.

Check it in 1D WIN-NMR:

Load the proton spectrum of glucose D:\NMRDATA\GLUCOSE\1D\H\GH\001999.1R. From the **Output** pull-down menu choose the **Define Plot** option. Click on the **x-Area...** button and set the **Start Value** and **Stop Value** so that the ring and methyl protons are displayed. Make sure that the **Auto Arrange to Paper** option is set. Close the dialog box and click on the **y-Area...** button. Change the **Offset** value to 20% and the **Y-Scale** to 120%. Close the dialog box and change the spectrum in the display using the x and y expansion buttons. Click on the **Show Area** button, the spectrum display will return to the plot region as defined before. Click on the **Plot** button. **The Page Layout** dialog box appears, click on the **Print** button to produce a plot. For further details regarding the **Page Layout** dialog box proceed to section 4.7.2. If problems arise with your plotting device check the entries in the **Printer Setup** dialog box discussed in the next section and/or follow the instructions givne in section 2.3 of this volume.

4.7.2 Page Layout

Depending on which display mode (*1D*, *Dual Display*, *Multi Display*) has been selected, different dialog boxes for **Page Layout** appear on screen:

4.7.2.1 Page Layout Dialog Box in Normal 1D Display Mode

From the **Output** pull-down menu choosing the **Page Layout** option opens the **Page Layout** dialog box (Fig. 4.16). The same dialog box may also be opened using the **Plot** button in the **Define Plot** submenu accessible in the the **Output** pull-down (section 4.7.1), or may be opened with a right mouse click on the **Print!** button in the left side button panel. The **Page Layout** dialog box allows you to control the previously set parameters or directly set the parameters that define a plot using the edit fields, to activate commands or to open additional dialog boxes. All these operations affect the appearance of the plot page.

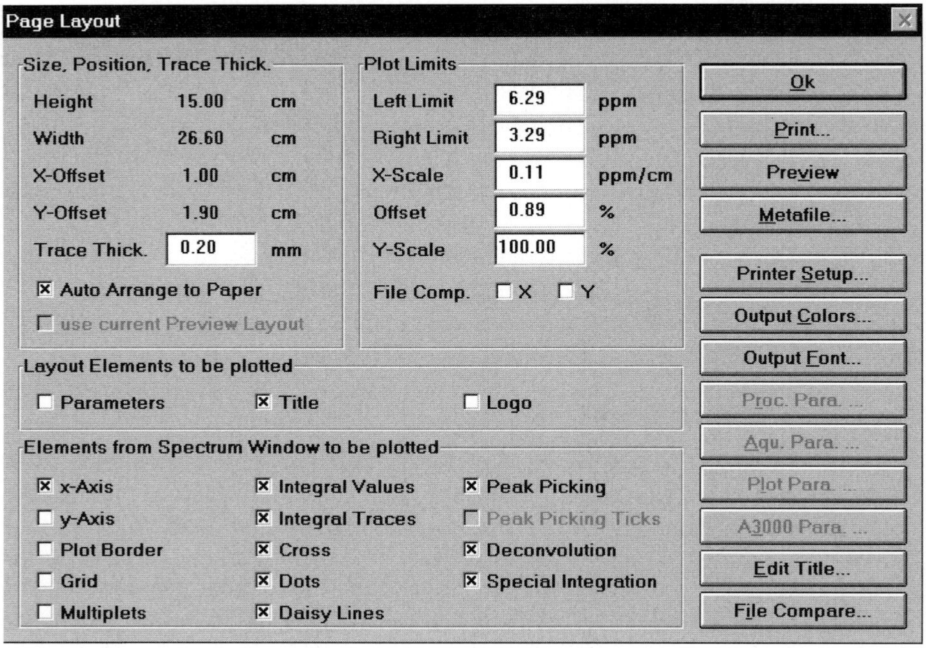

Fig. 4.16: **Page Layout** dialog box in normal *1D* display mode.

The use of the individual elements of the **Page Layout** dialog box is as follows:

Size, Position, Trace Thick. In this group box, you may set the dimensions and offsets of the plot and, for laser printers, the line widths used for hardcopy output.
Two **Auto Arrange** options exist to adjust the layout either to fit the plot paper size or to a given layout already existing in the *Preview* window. If one of these options is checked, some of the fields in this group can no longer be edited.

Plot Limits You may specify the limits of the spectrum to appear inside the plot area using the appropriate edit fields in this group box, provided that the **File Comp. X/Y** boxes are not checked. To plot the current spectrum over the same frequency range and in the same plot area as a previous plot, or to compare intensities in these plots, check the **File Comp. X** and **-Y** boxes respectively. Before either option can be selected, the reference spectrum must first be specified in the **File Compare** dialog box opened with the **File Compare...** button.

Layout Elements to be plotted This group box contains self-explanatory display options for the plot.

Elements from Spectrum Window to be plotted	Items that are currently active in the *Spectrum* window can be selected for the hardcopy or **Preview** output by checking the respective filters in the **Display Options** group.
Printer Setup	Clicking this button opens the **Printer Setup** dialog box through which the user can select the output device from a list of all devices known to the MS-WINDOWS operating environment.
Output Colors...	Selecting this button opens a dialog box in which the colors of different elements in the plot can be edited. Important: If you are using a monochrom output the **Output Colors...** option should be set to monochrom, this will ensure that all the selected objects appear in the hardcopy.
Output Font...	This button opens a dialog box in which you may select the font to be used in plotter or preview output for the spectrum window (scale, peak picking, integral). It does not affect the fonts for the parameter and the title window.
Proc. Para. ...	This button allows the selection of the processing parameters that will be included in the plot. The button is active only if **Parameters** has been selected in the **Layout Elements to be plotted** group box.
Aqu. Para. ...	This button works in the same way as the **Proc. Para...** button and allows the selection of the acquisition parameters to be included in the plot. The button is active only if **Parameters** has been selected in the **Layout Elements to be plotted** group box.
Plot Para. ...	This button works in the same way as the **Proc. Para...** button and allows the selection of the plot parameters to be included in the plot. These parameters are stored in a .PLT file and is the file referred to when the spectral comparison mode is selected in the **Plot Limits** group box using the **File Comp. X** and **File Comp. Y** check boxes.
A3000 Para. ...	Allows you to select all the parameters stored in a .DIS file. This file is created during the conversion of Aspect 3000 files (DISNMR and DISMSL files) into the WIN-NMR format.
Edit Title...	An edit field for a title is opened and the title may be stored in a .TIT file. Multi-line titles can be added to the plot if the **Title** box in the **Layout Elements to be plotted** group box is checked.
File Compare	Opens a **Select the File** dialog box in which you can select the filename of the reference spectrum to base the current plot limits and scaling on.
Print ...	Closes the **Page Layout** dialog box and opens the **Print** dialog box. As well as printing the page layout, this dialog box contains options to select and configure the output device (see section 4.7.4.)
Preview	Closes the **Page Layout** dialog box and passes the output into the *Preview* window for output control prior to plotting. **Preview** offers a variety of possibilities for further graphical processing which are discussed in section 4.7.3.

Metafile... Closes the **Page Layout** dialog box and opens a **Save Metafile** dialog box for storing the plot layout in the device independent MS-WINDOWS metafile format (<NAME>.WMF). Metafiles can be imported into word processing programs or graphics and desktop publishing applications for further manipulation. Metafiles can be read directly into the *Preview* window of 1D WIN-NMR.

Note: The fonts for the title and the parameter windows must be set interactively by double clicking the corresponding windows with your left mouse button which opens a corresponding dialog box.

Check it in 1D WIN-NMR:

Load the ^{1}H spectrum of peracetylated glucose D:\NMRDATA\GLUCOSE\1D\H\GH\002999.1R and try the rapid mode for plotting using the **Print!** button in the button panel. First click with your right mouse button on it to inspect the layout parameters in the appearing dialog box. Perform several plots with a few of the entries and options changed first and initialize the plot operation by clicking then with your left mouse button on the **Print!** button.

Check it in 1D WIN-NMR:

Set up a plot layout similar to the example given in Fig. 4.17. Load the ^{13}C spectrum of glucose D:\NMRDATA\GLUCOSE\1D\C\GC\001999.1R, check its calibration and perform a peak picking. From the **Output** pull-down menu choose the **Page Layout** option. Within the **Page Layout** dialog box set the **Left Limit** and **Right Limit** to 100 ppm and 10 ppm, the **Offset** to 10% and the **Y-Scale** to 200%.

In the box **Layout Elements to be plotted** choose **Parameters** and **Title**. Click on the **Edit Title...** button in the right side button panel to enter a title. Clicking on the appropriate buttons, define the acquisition, processing and plotting parameters that you require to be printed with your spectrum.

In the box **Elements from Spectrum Window to be plotted** choose the items **x-Axis** and **Peak Picking**.

Click the **Output Colors...** button and choose either the monochrom or any other color combination according to your output device.

Click the **Output Font...** button and define the output fonts for and the style according to your preferences. Note that this influcences only the fonts in the *Spectrum* window (peak picking values, scale (ppm/Hz) values), but not the fonts for the title and parameter windows of the plot layout. The latter fonts may be set with the **Preview** option (4.7.3).

Click on the **Preview** button and inspect your layout, if necessary modify the layout.

If necessary select the **Printer Setup...** button and setup your printing device.

Initiate the plot with the **Print ...** button in the **Page Layout** dialog box or from the **Printer Setup** dialog box.

If necessary use the **Help** tool for more detailed information.

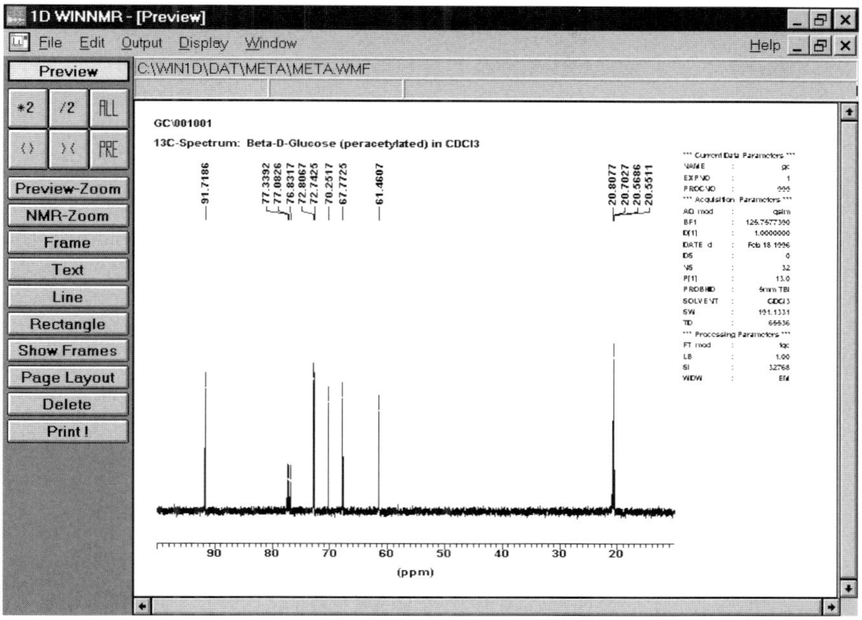

Fig. 4.17: Layout example in the *Preview* window.

Check it in 1D WIN-NMR:

Load the proton spectrum of glucose D:\NMRDATA\GLUCOSE\1D\H\001999.1R. Check its calibration. Choose the region of the spectrum containing the ring protons (6 – 3.5 ppm). In this region perform a peak picking in Hz and set/calibrate the integrals. From the **Output** pull-down menu select the **Page Layout** option to open the **Page Layout** dialog box. Check that the **Auto Arrange to Paper** option is set; inspect the plot limits, **Offset** and **Y-Scale** and adjust if necessary according to your preferences. Select **Parameters** and **Title** and **x-Axis** as **Layout Elements to be plotted** and create a title with the **Edit Title...** button. Choose in the box **Elements from Spectrum Window to be plotted**, **x-Axis**, **Integral Values**, **Integral Shapes** and **Peak Picking**. Clicking on the appropriate buttons, define the acquisition, processing and plotting parameters that you require to be printed with your spectrum. If necessary, set the output colors and output fonts by using the corresponding buttons. Click on the **Preview** button and inspect your layout, if necessary modify the layout. Finally plot the spectrum.

Check it with 1D WIN-NMR:

Use the same layout as above and store it as a metafile using the **Metafile...** button in the **Page Layout** dialog box. If you have a word processing program, e.g. WORD for WINDOWS running on your PC, try to import this metafile picture into a text file written with your word processing program.

Check it with 1D WIN-NMR:

To illustrate plotting a series of spectra with the same layout use the three ^{13}C DEPT spectra of glucose. Load the third of the three ^{13}C DEPT spectra of glucose D:\NMRDATA\GLUCOSE\1D\C\GCDP\003999.1R, which will be used as the reference spectrum. Reduce the y-scale and move the baseline to the middle of the display. Choose a window large enough to include both positive and negative peaks. Perform a peak picking and save the peak picking region. From the **Output** pull-down menu select the **Page Layout** option. Set the plot limits, **Left Limit** and **Right Limit**, to display the region of the ring carbons (100 – 50 ppm). Adjust the **Offset** to 50% and the **Y-Scale** to 200% to adapt the layout for both positive and negative peaks. Set the other layout options to include acquisition, processing and plot parameters, title and peak picking. Check your layout in the *Preview* window before plotting the spectrum. Save the spectrum using the extension 003001.

Now load the first DEPT spectrum. Using the third DEPT spectrum as the calibration file perform a peak picking. In the **Page Layout** dialog box check the x- and y- **File Compare** boxes and click on the **File Compare...** button to select the already plotted third DEPT spectrum as the calibration file. Click on the **Edit Title...** button to create an appropriate title. Click on the **Preview** button, accept clearing the *Preview* window and inspect the layout, which should now be the same as for the third DEPT spectrum. Plot the spectrum.

Repeat the process for the second DEPT spectrum. All three spectra should be plotted in exactly the same way.

4.7.2.2 Page Layout Dialog Box in the Dual and Multiple Display Mode

If your are in the *Dual* or *Multi Display* mode or if you are doing special processing with a dual display on screen, then selecting the **Page Layout** option in the **Output** pull-down menu opens a slightly different type of **Page Layout** dialog box (Fig. 4.18). Most of the edit fields are self-explanatory or have the same function as in the normal **Page Layout** dialog box.
The plot is not totally defined by the options available in the **Page Layout (Dual/Multi Display)** dialog box. Additional dual/multi display options can be set either by choosing **Options...** in the **Display** pull-down menu or clicking on the **Options...** button in the Dual-, Multi Display button panel.

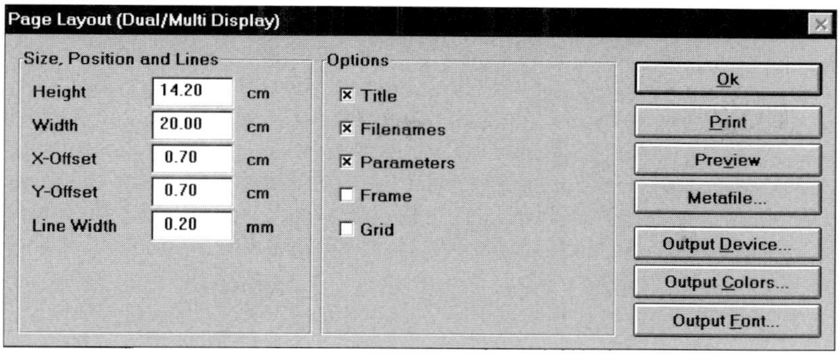

Fig. 4.18: **Page Layout** dialog box available with dual and multiple displays.

Check it in 1D WIN-NMR:

Load one of the proton NOE-difference spectra of glucose, e.g. D:\NMRDATA\GLUCOSE\1D\H\GHNO\1D\002999.1R. From the **Display** pull-down menu choose the **Dual Display** option. Select the reference spectrum D:\NMRDATA\GLUCOSE\1D\H\GHNO\1D\008999.1R as the second trace spectrum. Use the vertical scroll bar to adjust the position of the baseline of both spectra so that it is close to the bottom of the display window. Using the ***2** button in the main button panel, increase the vertical scale until the NOEs in the NOE difference spectrum are clearly visible. (Ignore the fact that the large negative signal in the NOE difference spectrum will be cut off). Click on the **Options...** button in the Dual Display button panel. In the **Dual Display Option** dialog box set the plot limits in the 'Region (Main Trace)' box to 6.2 ppm and 3.8 ppm. Make sure that **Y Axis Units** is set to **absolute** and that **Separate** is selected in the **Draw Option** box. In the second trace window adjust **Factor** to avoid any part of the second spectrum being cut off. Click the **OK** button to close the dialog box. From the **Output** pull-down menu choose the **Page Layout** option. Set the options in the **Size, Position and Lines** box according to your preferences. In the **Options** box select **Filenames** and **Parameters**. Check, and if necessary, adjust the output colors and the output fonts to suit your selected output device. Click on the **Preview** button to check your layout and to make any final adjustments prior to plotting. Experiment with the other options (**Title**, **Frame** and **Grid**) available in the **Option...** dialog box and plot the new page layouts.

Check it in 1D WIN-NMR:

From the series of selectively decoupled spectra of glucose load the reference spectrum D:\NMRDATA\GLUCOSE\1D\H\GHDC\001999.1R. From the **Display** pull-down menu choose the **Multiple Display** option. In the **Select Multi Display Mode** dialog box select all the seven spectra, starting with the reference spectrum. Click on the **OK** button to close the dialog box and initialize the multiple display. Click on the **File Param.** button and inspect the

NC values for each spectrum. Adjust the **Y-Scaling** value to compensate for different **NC** values, i.e. set **Y-Scaling** to 1 for a **NC** value of 2 and **Y-Scaling** to 8 for a **NC** value of 5. Click on the **OK** button to close the dialog box. Using the zoom mode button in the button panel and the **Mouse Grid** button, adjust the display window until the region of the ring protons is displayed to your satisfaction. From the **Output** pull-down menu choose the **Page Layout** option and select the same parameters, options, colors, fonts and the printer setup as used in the dual display above. Use the **Preview** option for a final inspection then plot this "Stack Plot" layout.

Check it in 1D WIN-NMR:

To illustrate the "Separate Plot" layout for a series of spectra load the same series of spectra making sure that the reference spectrum is the main trace in the **File Param**. dialog box. Click on the **Separate** button to initialize the multiple display which consists of 7 equal windows each containing a single spectrum with its own x-axis.

Adjust the vertical scale and the vertical offset for the main trace spectrum to obtain the best display of the signals for the ring protons. Click on the **File Param.** button and set the plot limits to 6.0 ppm and 3.5 ppm. Click on both the **F1/F2 for all** and the **Y for all** buttons to transfer the plotting information from the reference spectrum to the other six spectra. Click on the **OK** button to close the dialog box. Do not click on the **Return** button as this will lose all the "Separate Plot" plotting parameters that have just been set. From the **Output** pull-down menu choose the **Page Layout** option and select the same parameters, options, colors, fonts and the printer setup as used in the multiple display above. Use the **Preview** option for a final inspection before plotting this "Separate Plot" layout.

Check it in 1D WIN-NMR:

Using the NOE spectra in the directory D:\NMRDATA\GLUCOSE\1D\H\GHNO\1D\) design a "Separate Plot" layout. Use the various options available with the Multi Display buttons to set the baseline of all the NOE difference spectra and the reference spectrum close to the bottom. (Again the large negative signals in the NOE difference spectra will be cut off but ignore this fact.) Uniformly adjust the vertical scale of the NOE difference spectra to the largest NOE. Choose the best layout for the reference spectrum. Plot this series of spectra.

4.7.3 Preview

The *Preview* window performs a number of different tasks and offers a wide variety of possibilities to arrange your final layout. Normally the layout is composed of your spectrum and/or expansions of it, different parameter lists, text files and graphical elements such as structural formulae, designed within or imported into the *Preview* window. It also allows layout parameters such as colors, fonts and style of textfile characters to be defined and layouts (including only the individual frames) to be saved for use in further applications. In view of this variety of layout options and the fact that everyone has their own preferences this section is restricted to a short overview. The setup of typical layouts will be demonstrated and you may try them out with a few *Check its*. It is recommended, however, that you explore these preview options following the instructions given in the corresponding **Help** menus.

The *Preview* window may be opened directly with the **Preview!** button in the button panel or via the **Preview** option in the **Output** pull-down menu. If the *Preview* window is not empty a corresponding message and the question to clear it appears on screen. In both cases the current display is passed directly into the *Preview* window for further graphical processing according to the actual page layout and preview parameters. Clicking with the right mouse button on the **Preview!** button allows you to first inspect and adjust the 'Page Layout' parameters prior to the import of the spectrum. With the **Preview** window option in the main **Window** pull-down menu on the other hand you simply switch to the *Preview* window without updating it with the actual display in the *Spectrum* window.

The *Preview* window has its own set of dedicated pull-down menus (**File, Edit, Output, Display, Window**) as well as its own button panel.

For setting up the final layout it is often convenient to choose a dual *Spectrum/Preview* window display mode (Fig. 4.19). This is accomplished by choosing the **Spectrum/Preview** option in the **Window** pull-down menu of 1D WIN-NMR.

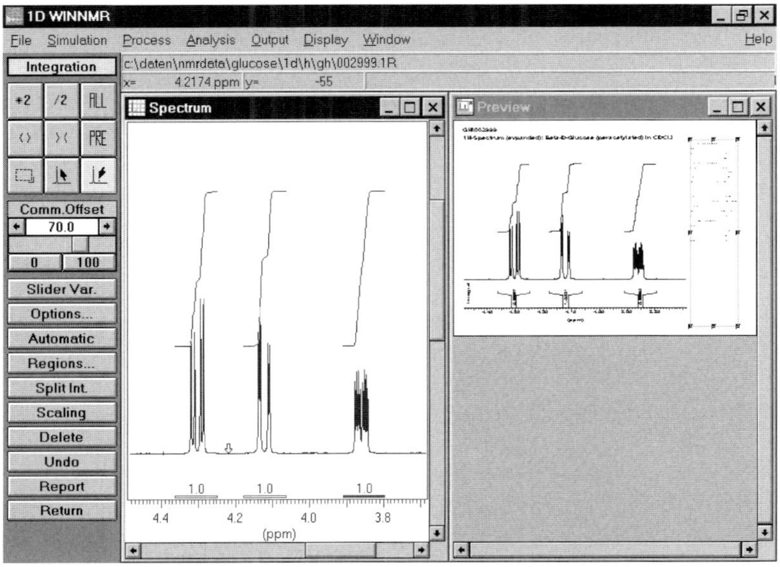

Fig. 4.19: Dual window display (*Spectrum/Preview*) with 1D WIN-NMR.

The *Preview* window has three main tasks:

- **Output control** – Before sending output to the hardcopy device (e.g. printer) the data can be previewed on the screen. The *Preview* window operates on a the basis of "What You See Is What You Get" (WYSIWYG) on the hardcopy device.
- **Merge function** – The *Preview* window allows you to merge together any stored output from different 1D WIN-NMR and/or 2D WIN-NMR functions, before sending the combined output to the hardcopy device. Such items include different spectra, peak picking lists or pulse programs and graphics elements e.g. structural formulae or the contents of the WINDOWS clipboard (Pasting option).
- **Edit function** – The size and position of all the graphical elements shown in the *Preview* window (spectra, listings, title, structure formulae etc.) and their colors may be modified. Simple graphics elements such as lines or rectangles may be created and added and different types of fonts for text and lists are available.

The *Preview* window offers two different operation modes:

- **Mouse Select Mode** – If none of the buttons in the button panel are highlighted, the mouse pointer is in *Select* mode, which allows you to select/deselect, edit, resize, drag and delete graphical elements. Graphical elements include spectra, text files, metafiles etc. Double clicking on a graphical element opens a dialog which contains a wide variety of options for manipulating the selected element.
- **Button Panel** – A series of buttons in the button panel allow you to open additional frames, to zoom or delete frames, to add and edit text files, lines or rectangles, to switch to the **Page Layout** dialog box and to initialize printing.

Check it in 1D WIN-NMR:

Load the proton spectrum of glucose D:\NMRDATA\GLUCOSE\1D\H\GH\001999.1R and use the **Preview!** button in the button panel for a rapid layout setup. Exploit the right mouse button option for first inspecting and adjusting the layout parameters, before the actual display in the *Spectrum* window is passed into the *Preview* window. Generate several layouts changing the layout parameters accordingly.

Check it in 1D WIN-NMR:

Load the proton spectrum of glucose D:\NMRDATA\GLUCOSE\1D\H\GH\001999.1R and set up a first layout containing the spectrum in the range of the glucose signals (6 – 1.5 ppm), the x-axis, the parameters, and a title. Use the **Page Layout** option in the **Output** pull-down menu and the corresponding dialog box for this purpose. Click on the **Preview** button to switch to the *Preview* window and inspect this basic layout. Hit the **Show Frames** button, select the title frame with the mouse cursor set inside this frame and a single mouse click. The title frame is marked now with eight "handles" – small black squares. Activate the corresponding **Textfile Options** dialog box, with the mouse cursor (+) still inside the text frame by double clicking your left mouse button and adjust the fonts if necessary. Proceed in the same way to adjust the fonts of the parameter frame.

From the **Window** pull-down menu choose the **Preview/Spectrum** option. Move back to the *Spectrum* window of 1D WIN-NMR and load the second spectrum D:\NMRDATA\GLUCOSE\1D\H\GH\002999.1R. Set up a second layout with a spectral region 5.8 – 5.0 ppm, including peak picking in Hz and the integrals. Inspect this layout in the *Spectrum* window and when you are satisfied store the spectrum as D:\NMRDATA\GLUCOSE\1D\H\GH\002001.1R. Do not change the layout in the *Spectrum* window at this stage. With the dual (*Spectrum/Preview*) window still on your screen move to the *Preview* window again; open a second frame by first clicking the **Frame** button. Make sure that this is a selected frame, marked again with eight "handles" – small black squares. Double click the left button of your mouse to open the dialog box of this second frame. Click on the **Read NMR-File** button to read in the file D:\NMRDATA\GLUCOSE\1D\H\GH\002001.1R stored before. Click on the **Options** button and in the dialog box select the **x-axis**, **Line**, **Integral Shape**, **Integral Value** and the **Peaks** (picking) options before clicking the **OK** button. The *Preview* window now shows a superposition of both layouts. Readjust the size and the position of this second frame for the best layout and modify additional layout parameters in the corresponding **Options** dialog boxes, opened with a double mouse click in the correspondingly activated frames.

Now open a third smaller window and load the meta-graphics file D:NMRDATA\GLUCOSE\GLUCOSE.WMF containing the molecular formula of peracetylated β-D-glucose.

Resize and rearrange the two additional frames for the best overall layout (Fig. 4.20) For this purpose activate the frame and modify size and position by using the drag and move functions of your mouse.

Try out **the Preview-Zoom** and the **NMR-Zoom** buttons. The **Preview Zoom** allows you to zoom any part of the layout in a reversible way. Use the **All** button in the button panel to re-establish the previous layout. The **NMR-Zoom** is used in conjunction with the **Frame** button to allow the rapid display of expanded regions of the spectrum in additional frames.

Check it in 1D WIN-NMR:

Using your own ideas and preferences set up layouts for the 1D ^{1}H and ^{13}C spectra in the directories GH and GC, GCDP, GCJM respectively. These 1D spectra will serve for documentary purposes, as well as reference data for the elucidation of the unknown structure of the peracetylated oligosaccharide as discussed in chapter 6.

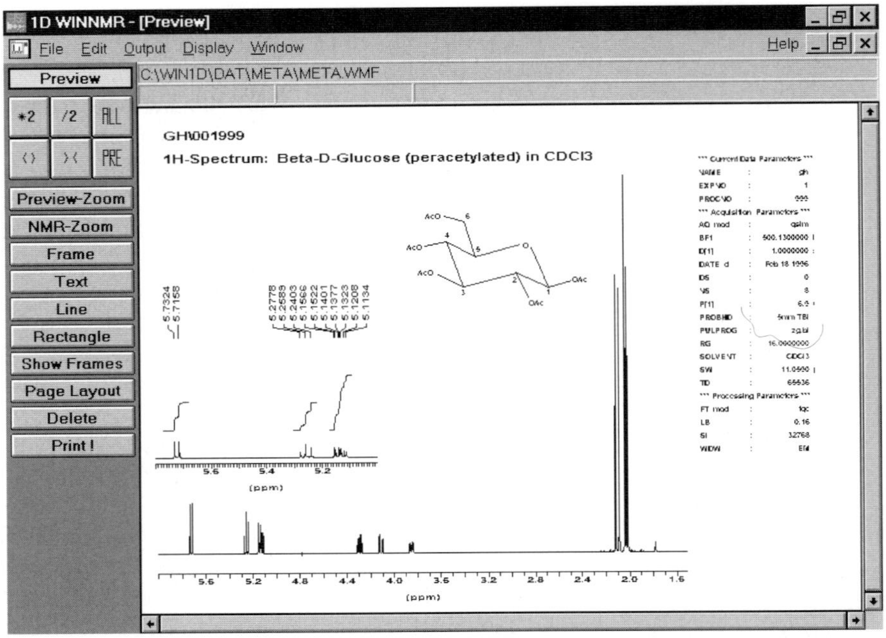

Fig. 4.20: Layout example

Check it in 1D WIN-NMR:

Load the first ^{13}C T_1 spectrum of glucose D:\NMRDATA\GLUCOSE\1D\C\GCT1\1D\001999.1R. Set up the acquisition and processing parameters of this T_1 inversion recovery experiment with the corresponding options in the **Output** pull-down menu. Use a dual window (*Spectrum/Text*) for this purpose. After selecting the required entries click on the **Edit** button to transfer the entries into the text editor. Store this textfile as T1PAR.ASC. Set up a multiple display layout, as previously described; include all the spectra in the series and display the carbonyl peaks region of the peracetylated glucose in the „3D" mode. Click the **Page Layout** button in the **Output** pull-down menu and adjust the various multidisplay parameters (**Size**, **Position**, **Trace Thick.)** and options according to your own ideas. Click on the **Preview** button to move this display into the *Preview* window. Use the dual window display (*Preview/Spectrum*) to facilitate the setup of the final layout.

Click on the **Frame** button and open a second large frame in the *Preview* window. Double click the left button on the mouse to open the dialog box and click on **Read NMR-File...** to read in the normal ^{13}C spectrum of glucose D:\NMRDATA\GLUCOSE\1D\C\GC\001999.1R. Adjust its size and set the various plot options for this spectrum. Finally readjust the sizes of both windows for the best layout. Open two additional windows and read in the previously saved parameter textfile and the glucose structural formula respectively. Use the **Preview Zoom** function to inspect the individual windows in detail. Use the **Line** option to graphically connect the multiple display with the corresponding spectral region of the normal ^{13}C spectrum.

Check it in 1D WIN-NMR:

Plot the other series of ^{1}H spectra obtained with the selective decoupling (GHDC), NOE (GHNOE\1D), ROE (GHRO) and TOCSY (GHTO) experiments. Using either the dual or the multiple display options set up the plot layouts according to your own preferences. Again, this series of 1D spectra will serve for documentary purposes, as well as reference data for the elucidation of the unknown structure of the peracetylated oligosaccharide as discussed in chapter 6.

Additional functions available within the **Output** pull-down menu, or accessible from within the **Page Layout** or the **Preview** dialog boxes by clicking the appropriate buttons are as follows:

4.7.4 Printer Setup... , Print...

Printer Setup... allows you to select and configure the output device from a list of all devices known to the MS-WINDOWS operating environment. **Print...** opens the same dialog box, but includes the plot command.

4.7.5 Copy

Sends the plot as a vector-oriented metafile to the MS-WINDOWS clipboard from where it can be transferred into any other MS-WINDOWS program. The plot area of the output is taken from the currently displayed area in the *Spectrum* window. Parameter lists and the title cannot be transferred using this method. They can be copied by transferring the output, including the title and lists, first into the *Preview* window, from where it may be copied to the MS-WINDOWS clipboard using the **Copy** option in the **Edit** pull-down menu in the *Preview* window menu bar.

Check it in 1D WIN-NMR:

Set up a layout including a spectrum, a textfile and a graphic file with the structural formula of glucose and try to copy this layout via the clipboard into your word processor program, e.g. WORD for WINDOWS.

4.7.6 Metafile...

In the **Output** pull-down menu of 1D WIN-NMR, **Metafile...** will store the selected plot in a MS-WINDOWS metafile. Metafiles may also be stored or loaded from a corresponding frame in the *Preview* window with the **Save Metafile** and **Open Metafile** options in the **File** pull-down menu of the *Preview* window.

Check it in 1D WIN-NMR:

Again set up a layout including a spectrum, a textfile and a graphic file with the structural formula of glucose and save the preview display as a metafile. Try to paste this file into your word processor program, e.g. WORD.

4.7.7 ACQ., PROC., PLOT and A3000 Parameters

For NMR data measured on UXNMR/XWINNMR based spectrometers, these options opens the appropriate dialog box showing the ACQuisition (AQS), the PROCessing (FQS) and the PLOT (PLT) parameters. For NMR data files measured on DISNMR/DISMSL based spectrometers (AM, AC) the corresponding parameters are accessible in the **A3000 Parameters**. Parameters may be edited, printed or transferred to the *Preview* window, from where they may be stored as a textfile.

4.7.8 Title...

Lets you edit the title of the spectrum, send it to the *Preview* window and store it from there as a title file (<NAME>.TIT).

4.7.9 Pulse Program..., AU Program...

Pulse Program... lets you view the file that contains the pulse program for the experiment while **AU Program...** lets you view the automation program for the spectrum. To work correctly, the corresponding files must exist in the PC (pulse program) and the AU (AU program) directory that was specified at program installation (see **Help** for installation).

Check it in 1D WIN-NMR:

Load the ^{13}C DEPT spectrum D:\NMRDATA\GLUCOSE\1D\C\GCDP\001999.R and try out the **Pulse Program...** button, available in the **Output** pull-down menu, to inspect the DEPT pulse program.

4.7.10 History...

Lets you view the history file, which contains all the commands for the current processing session that have changed the data in the *Spectrum* window.

4.8 Display of 2D Spectra with 2D WIN-NMR

4.8.1 Buttons with 2D WIN-NMR

This section describes the function of the most important buttons in the button panel (Fig. 4.21). Further options available within the **Display** pull-down menu of 2D WIN-NMR will be discussed in section 4.8.3. Two procedures to set contour levels for a 2D spectrum displayed as a contour plot are discussed in section 4.8.2.
In many cases the symbols on most of the buttons are self-explanatory. All buttons are disabled when the program is started and enabled after spectral data has been loaded. The panel remains disabled if a raw (SER) data file is opened. Use the **Help** tool for detailed information concerning the various options and how to use them.

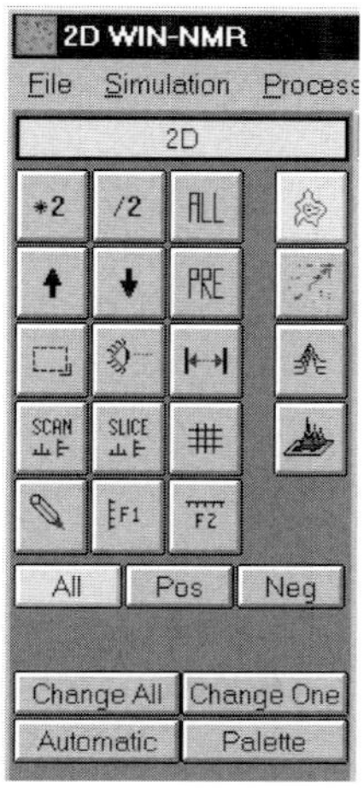

Fig. 4.21: 2D WIN-NMR display button panel with the *Contour* display mode activated

Four different display modes for 2D spectra are available:

Contour mode: The currently selected 2D spectrum is displayed as a contour plot, the usual mode of display. At the same time corresponding local mode buttons appear in the button panel (Fig. 4.21) which allow to set contour levels. See section 4.8.2 or use the **Help** tool for more information concerning these options.

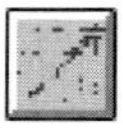

Density mode: Displays the 2D data set in the less common density plot representation.
Local functional buttons (**Linear**, **Log.**, **Intens. Rng**, **Threshold**, **Palette**) appear in the panel for manipulation of the representation according to your requirements. Use the **Help** tool for more informations.

Stacked mode: Displays the data set in a stacked plot form. Local functional

buttons (**Manual**, **Grid**, **Cutting Level**) appear in the panel for manipulation of the representation according to your requirements. Use the **Help** tool for more informations.

3D mode: Causes the data set to be displayed in *3D* mode. A new local button block appears when this mode is activated, which allows you to control the way in which the 3D plot is displayed (**Normal/Transparent**, **Standard/Inverted**) and the direction from which the 3D plot may be examined (**Front Sight/Back Sight**, **Right Sight/Left Sight**). Furthermore a cutting level (**Cutting Level**) may be set, which cuts intense peaks out allowing the closer examination of small peaks. Use the **Help** tool for more informations.

The 15 buttons arranged blockwise in the button panel are denoted as the main operation buttons. They are permanently present in the button panel and can be used to perform the desired operation on the selected data set:

Scaling buttons: In the *Contour* mode ***2** increases and **/2** decreases, the vertical scaling of both the projections and of the contours by a factor of 2. To individually scale only the projection or only the contours, use the menu item **Select** in the **Layout** submenu of the **Display** pull-down menu.

In the *Contour* mode these buttons move the projections up/down. In the *Stacked*- and *3-D* mode these buttons move up/down the displayed region.

ALL: The whole data set is displayed on the screen.

PRE: Return to the previously displayed part of the spectrum.

Expand: Causes a dialog box to appear on screen, which allows you to define the regions in both dimensions of a 2D spectrum, either in ppm or Hz. For quick expansions the mouse may be used within the spectrum window.

View: Activates the display of the pointer co-ordinates (ppm, Hz, pts) when the mouse pointer is moved in to the contour plot.

Distance: Allows you to calculate the distance (in ppm, Hz or data points) between two points of the 2D-contour plot.

Scan: A cross-hair cursor appears on the screen, which allows you to examine the spectrum slices along the horizontal and vertical bars. Local mode buttons **Columns** and **Rows** appear to scan either columns or rows. A selected row or column can be "frozen" on the screen so that peak intensities can be measured or contour levels can be adjusted (see section 4.8.2).

Slices: Select either a row or a column by its row-/column number and transfer it to 1D WIN-NMR for further processing.

Grid: Causes a grid to appear on the spectrum which facilitates subsequent spectral analysis.

Annotate: The type (text, intensity, ppm, Hz or points) and the font, type and color for annotation may be defined.

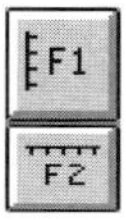

Change Axis: The axis units can be toggled between ppm, Hz and points for the F1 and F2 axis respectively.

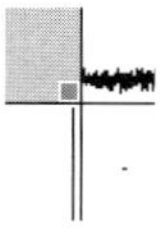

Projection size box: Whenever projections have been selected this small box is located at the upper left corner of the 2D spectrum. Dragging this box simultaneously changes the relative size of the projection fields and the 2D spectrum display field.

Check it in 2D WIN-NMR:

Load the 2D COSY (magnitude mode) spectrum of peracetylated glucose D:\NMRDATA\GLUCOSE\2D\HH\GHHCO\001999.RR. The 2D spectrum appears in the default contour display mode and shows only positive cross peaks. Use the mouse to select a suitable expansion.

Try out the effect of the ***2**, **/2**, **ALL** and **PRE** buttons in the button panel.

Use the **Change Axis** button to change the axis units of your 2D spectrum.

Click the **Expand** button and define an expansion by entering the appropriate ppm values for F1 and F2. Try out the **View** and **Distance** buttons.

Now use the **Scan** function to inspect the rows and columns of your 2D spectrum. Note the number of one of them and use the **Slice** function to transfer it to 1D WIN-NMR for further inspection.

Try out the effect of the **Grid** and the **Change Axis** buttons.

Check it in 2D WIN-NMR:

Use the same 2D spectrum and check the other display modes (*Density* mode, *Stacked* mode and *3D* mode). To speed up your work first select a suitable expansion. Try out the effect of the correspondingly available buttons in the button panel: **Linear**, **Log.**, **Intens. Rng**, **Threshold**, **Palette** with the *Density* mode, **Manual**, **Grid**, **Cutting Level** with the *Stacked* mode and

Normal, **Transp.**, **Standard**, **Inverted**, **Front-**, **Left-**, **Back-**, **Right-Side**, **Cutting Level** with the *3D* mode.

4.8.2 Setting Contour Levels

With a 2D spectrum displayed in the *Contour* mode, the number, intensitiy and color of contour levels have to be defined. Two procedures to set contours exist:

- Most usually contour levels are set using the buttons in the button panel correspondingly accessible with the *Contour* mode (Fig. 4.21). **All** displays both positive and negative cross peaks, **Pos** displays only positive cross peaks and **Neg** displays only negative cross peaks. With the buttons **Change All**, **Change One**, **Automatic** and **Palette** corresponding dialog boxes appear and allow you to adjust the value, the number, the distance and the color of the contour levels according to your preferences, either interactively or automatically. Please consult the **Help** tool for more informations if necessary.
- As an alternative contour levels may be defined using suitable rows or columns of a 2D spectrum. This is especially useful if very weak cross peaks with intensities close to the noise level should be detected. To set levels in this way the *Scan* mode is activated and the option to "freeze" a suitable row or column is exploited.

Please follow the instructions given in the next *Check its* to become familiar with these two procedures.

Check it in 2D WIN-NMR:

Load the 2D COSY spectrum (magnitude mode) of peracetylated glucose D:\NMRDATA\GLUCOSE\2D\HH\GHHCO\001999.RR and click on the **Change One** button to adjust the value of the lowest level. Select the level field in the upper right corner (lowest level) and adjust its value with the scroll bar, click the **OK** button. Now with the lowest level set click the **Change All** button, select **Options>>** and choose 4 levels to be displayed. Choose the ⇧ option, vary the **Factor** value and try out the **Increment/Multiply** options to get the best presentation. Select 6 levels for display and adjust the various parameters again. To change the colors of the individual contours, from the **Display** pull-down menu choose the **Colors...** option. Select different colors for the six levels and inspect the display.

Test the **Automatic** function for automatic calculation of 6 levels and try out several values for **Factor**.

Check it in 2D WIN-NMR:

Load the phase sensitive, double quantum filtered 2D COSY spectrum of peracetylated glucose D:\NMRDATA\GLUCOSE\2D\HH\GHHCODF\ 001999. RR. This 2D spectrum contains both positive and negative peaks and again appears as a contour display spectrum. To speed up the display refresh time use the **Expand** button in the button panel to select a suitable expansion. To differentiate between positive and negative levels on your display, first choose the **Colors...** option in the **Display** pull-down menu and set the first three levels (positive) to black and the residual three levels (negative) to red. Use the **Change One** button to adjust the value of the lowest positive level. Select the level field in the upper right corner, adjust its value with the scroll bar and click the **OK** button. Finally use the **Change All** button, select **Options>>** and choose the ⇧ and *+/–* (to display positive and negative levels) options and totally 6 levels to be displayed. Vary the **Factor** value and try out the **Increment/Multiply** option to get the best presentation. Try out the **All**, **Pos.** and **Neg.** buttons on the button panel.

Test the **Automatic** function for automatic calculation of 6 levels and try out several values for **Factor**.

Check it in 2D WIN-NMR:

Load the heteronuclear magnitude mode 2D COSY spectrum of peracetylated glucose D:\NMRDATA\GLUCOSE\2D\CH\GCHCO\001999.RR. Hit the **Scan** button in the left side button panel and select a column showing the cross peaks of the methylene carbon, i.e. column 905 at 61.5647 ppm. "Freeze" this column by pressing the **Shift-Key** and the **left mouse button** simultaneously. Adjust the intensity of this column by clicking the left or right mouse button. Enter now the *Set Levels* mode by pressing simultaneously the **Control-** and the **Shift-Keys** and click on the left mouse button. An info text appears on the inlet and the new **Reset Levels** button is displayed. Position the mouse pointer to the desired position (the current intensity is monitored in the info inlet) and click the left mouse button. The selected level is displayed on the screen. Define six levels in this manner. If you want to cancel the selected levels click on the **Reset Levels** button. The system will leave the *Set Levels* and the *"frozen"* modes, will return to the standard *Scan* mode and you have to repeat the steps described above to define another set of levels. If you are satisfied with your selection exit the *Set Levels* mode by pressing again both the **Control-** and **Shift-Keys** simultaneously and by clicking the left mouse button. The new contours are calulated and displayed in the window and the system returns to the standard *Scan* mode.

4.8.3 Additional Display Options with 2D WIN-NMR

In addition to the display functions offered in the buttons panel, the **Display** pull-down menu (Fig. 4.22) contains additional options for controlling the display of the 2D matrix and its related spectral elements. It is possible to display standard 1D spectra, projections or slices on the F1 or F2 axis of a 2D plot together with the 2D spectrum. It is also possible to create additional expansion windows which can be moved, resized and deleted.

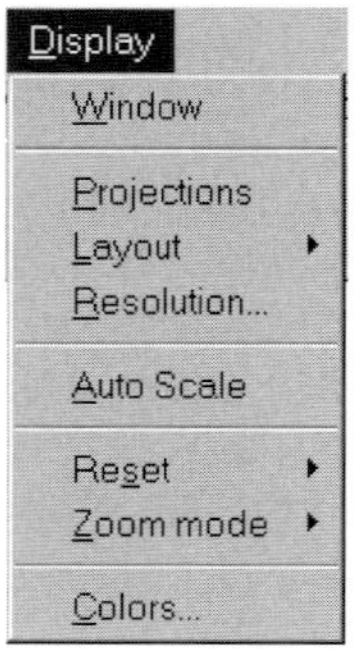

Fig. 4.22: The **Display** pull-down menu

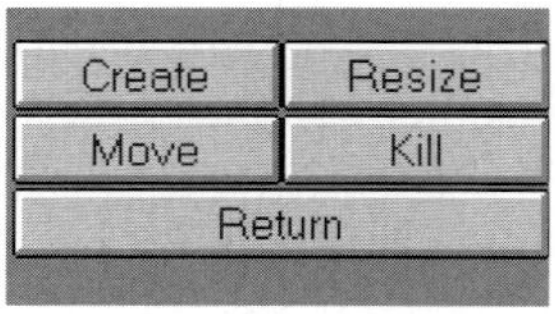

Fig. 4.23: Local operation button block available with the **Window** command

Window A local operation button block appears, which allows you to create, resize, move and delete additional expansion windows (Fig. 4.23). These expansion windows may be superimposed on the main window and enable you to inspect zoomed parts of the main window.

Projections Projections are 1D spectra calculated from the corresponding 2D spectrum. They may be calculated in different ways but their digital resolution is determined by the resolution of the 2D spectrum and is usually not as good as for a standard 1D spectrum. If available, a separately measured high resolution 1D spectrum should be used as a "projection" (see below). After being transferred from other systems, 2D data files initially have no calculated projection files and therefore will be displayed by default without projections along both the left (F1) and the top (F2) axis.With the **Projections** option several buttons to define regions and to calculate different projections along F1 and F2 appear. Refer to the **Help** routine for information on how to use these options.

Layout This command opens a menu with several options, which allow you to define the type of projection (**Projection**, **Spectrum...**, **Slice**, ...), to select the elements (**Select...**) that may be adjusted using the ***2**, **/2** buttons and to normalize your 2D spectrum (**Normalise**). The most common and recommended type of projection (**Spectrum...**) opens a dialog box (Fig. 4.24) where the filenames of the standard 1D spectra for F1 and F2 may be selected from the directory.

Resolution... Allows you to define the size of the actual 2D spectrum for display. Normally 2D data sets are much larger than can be held in internal memory for display and the data must be reduced (compressed) to fit into the internal data tables. The **Resolution** command allows you to choose what the resolution will be during the next **File/Open** operation. Values are 128x128, 256x256 or 512x512 points.

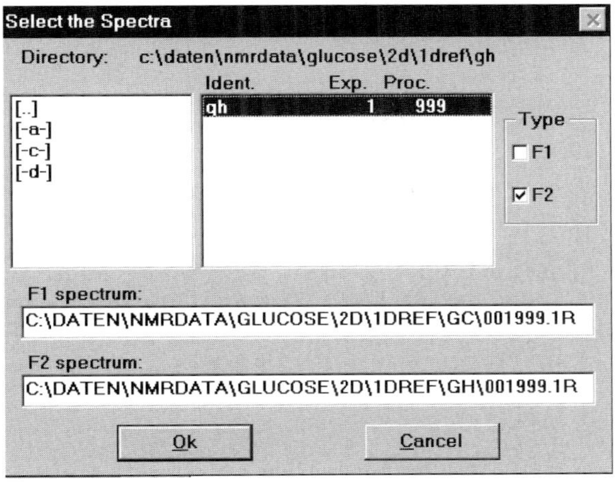

Fig. 4.24: The **Select the Spectra** dialog box available via the **Spectrum...** option in the **Layout** menu.

Auto Scale This command recalculates the extreme values on occasions where a data set is loaded and the values for the maximum/minimum data points are not correct.

Reset Opens a menu of options for resetting scaling factors in the display modes *Stacked* and *3D* for resetting scaled, shifted or expanded projections and slices.

Zoom Mode Two options are available for zooming your 2D spectrum. **Normal** allows zooming using the standard "rubber box" method while **Units** uses predefined increments.

Colors... Allows you to set the screen colors (contour levels, background).

Use the **Help** tool for additional informations.

Check it in 2D WIN-NMR:

Load the phase sensitive 2D COSY spectrum of peracetylated glucose D:\NMRDATA\GLUCOSE\2D\HH\GHHCODF\001999.RR. From the **Display** pull-down menu choose the **Layout** item which opens a dialog box. Select the option **Spectrum...**, which opens a further dialog box. Define the separately measured 1D spectrum D:\NMRDATA\GLUCOSE\2D\1DREF\GH\001999.1R) as the projection spectrum for both the F1 and F2 dimension. Click with your mouse cursor on the small grey rectangle in the upper left corner of your 2D spectrum (Projection size box). Holding the left mouse button down drag the box and resize your 2D spectrum and the 1D projections accordingly.

In the **Layout** menu choose the **Select...** option and define in the dialog box, which elements of your display **F1**, **F2** (projection) and **Contour** (2D spectrum) should be affected by the ***2**, **/2** and the ↑, ↓ buttons on the button panel and test these settings.

Use the **Help** tool for more information.

Check it in 2D WIN-NMR:

Use the same 2D spectrum to try out the other types of projections. From the **Display** pull-down menu choose the **Layout** menu and select **Slice** to interactively inspect slices along F1 and F2 and to define a suitable pair as projections. Do the same with the **Slice numerical** option.

Check it in 2D WIN-NMR:

Use again the same 2D spectrum and from the **Display** pull-down menu choose the **Resolution...** option. Reduce the resolution to 128x128 and load the 2D file again to make the size reduction current. The display build-up will be faster, but the resolution is obviously worse.

Check it in 2D WIN-NMR:

Use the same 2D spectrum to become familiar with all the **Window** options available in the **Display** pull-down menu. Display the full 2D spectrum at normal size (512x512). Click on the **Create** button and in an empty part of the display create a window. Now select a region you want to expand and copy it into the newly created window. Create a second window and expand a second part of your 2D spectrum. Use the **Help** tool for further information on how to do this and to display the axis within these windows.

4.9 Basic Processing Steps with 2D Spectra

Assuming that your 2D NMR data has already been processed, i.e. the Fourier transformations have been performed, the phases of the signals have been correctly adjusted to pure absorption in both dimensions if necessary and your 2D spectrum is stored on your PC's disk, there are still a few final processing steps to be performed, before the final layout is composed and plotting is initialized.

4.9.1 Calibration

Calibration is accessed via the **Manual Calibration** option in the **Analysis** pull-down menu (Fig. 4.25). This causes a cross-hair cursor to appear on the screen and the system waits for the setting of a calibration point (Fig. 4.26).

In most cases 1D spectra have already been measured before starting the 2D experiment and the spectral windows for the acquisition of the 2D spectrum are adjusted to encompass only the signals of the compound under investigation. As a consequence the signals of the reference compound, e.g. TMS, are usually outside of the spectral window and, are missing in the 2D spectrum.

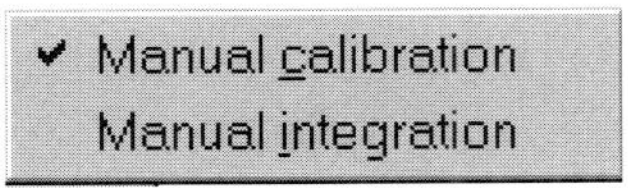

Fig. 4.25: **Analysis** pull- down menu.

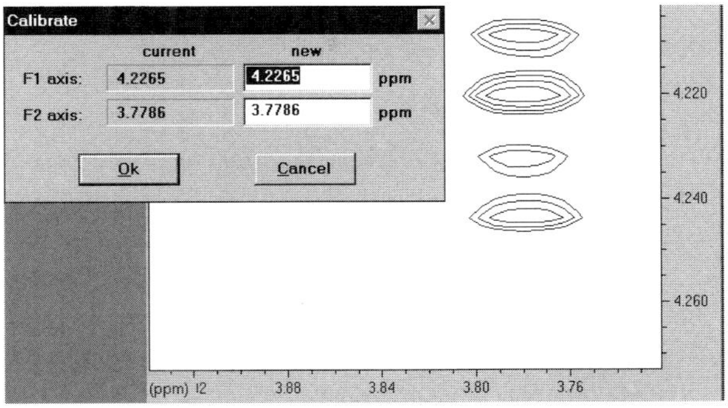

Fig. 4.26: **Calibrate** dialog box.

Therefore the calibration of 2D spectra is usually performed with respect to these 1D spectra, which then also serve as the projection spectra. The calibration of homonuclear 2D spectra only require one 1D spectrum whilst heteronuclear 2D spectra require two. In most cases the simplest procedure for calibrating 2D spectra is as follows:

1. Calibrate the corresponding 1D spectra in 1D WIN-NMR, store the calibrated 1D spectra and note the chemical shift value of a particular signal, which also appears in the 2D spectrum.
2. Enter 2D WIN-NMR and select an expansion showing the diagonal peak (in the homonuclear case) or the cross peak (in the heteronuclear case) of the same signal and calibrate it accordingly with respect to F1 and F2 using **Manual Calibration** in the **Analysis** pull-down menu.
3. Use the stored 1D spectra as projections and verify the correct calibration of the 2D spectrum. The selected peak(s) in the 1D projection spectrum should appear at the same position (ppm) as the corresponding cross peak in the 2D contour spectrum.

Check it in 2D WIN-NMR:

Load the 1D ^{1}H spectrum of peracetylated glucose D:\NMRDATA\GLUCOSE\1D\H\GH\ 001999.1R) and calibrate it with respect to TMS if this has not yet been done. Note the ^{1}H chemical shift of the anomeric proton.Now load the 2D COSY spectrum of peracetylated glucose D:\NMRDATA\GLUCOSE\2D\HH\GHHCO\001999.RR and choose an ex-pansion showing the diagonal peak of the anomeric proton to make calibration more precise. From the **Analysis** pull-down menu choose the **Manual Calibration** option. Move the cursor into the 2D spectrum and calibrate the diagonal peak of the anomeric proton. Use the value determined in the 1D spectrum for the F1 and F2 calibration. From the **Display** pull-down menu choose the **Layout** sub-menu and then the **Spectrum** command to set-up a display with the above 1D spectrum as the projection spectrum in both dimensions. Verify the correct calibration, then click on the **ALL** button to show the full 2D spectrum with its projections

Check it in 2D WIN-NMR:

To calibrate a heteronuclear 2D spectrum load the 1D ^{13}C spectrum of peracetylated glucose D:\NMRDATA\GLUCOSE\1D\C\GC\001999.1R) and calibrate it with respect to TMS, if this has not yet been done. Note the chemical shift of the anomeric carbon that appears at highest frequency.

Now load the 2D $^{13}C/^{1}H$ COSY spectrum of peracetylated glucose D:\NMRDATA\GLUCOSE\2D\CH\GCHCO\001999.RR and choose an expanison showing the cross peak of the anomeric proton/carbon to make calibration more precise. Select the **Manual Calibration** option in the **Analysis** pull-down menu. Move the cursor into the 2D spectrum and calibrate the cross peak connecting the proton and the carbon resonances of the anomeric methine group. Use the values determined in the appropriate 1D spectra for the F1 and F2 calibration. From the **Display** pull-down menu choose the **Layout** sub-menu and then the **Spectrum** command to set-up a display with the corresponding 1D spectra as the projection spectrua in the appropriate dimension. Verify the correct calibration and click on the **ALL** button to show the full 2D spectrum with its projections.

Check it in 2D WIN-NMR:

Calibrate all the other 2D spectra of peracetylated glucose in the directories D:\NMNRDATA\ GLUCOSE\2D\HH, ...CH in the same way.

4.9.2 Peak Picking

Peak picking in 2D spectra is accomplished using the annotation button in the button panel. Clicking this button opens a dialog box (Fig. 4.27). Cross and diagonal peaks may be annotated in ppm, Hz or points or may be annotated with a short text string (e.g. to indicate the corresponding nuclei numbers). This kind of annotation within the 2D spectrum must be made at this stage if it is to be included in the final plot layout.

Check it in 2D WIN-NMR:

To perform a 2D peak picking load the 2D $^{13}C/^{1}H$ COSY spectrum of peracetylated glucose D:\NMRDATA\GLUCOSE\2D\CH\GCHCO\001999.RR and expand a region where you want to perform a peak picking. Hit the button for annotation, select in the dialog box (Fig. 4.27) appearing on the screen ppm as the **Type** and define the **Color**, **Font**, **Style** and **Size** of the characters. Hit the **OK** button in this dialog box and perform the peak picking for a few cross peaks. Try out the various options available with the additional annotation buttons and annotate horizontally and vertically.

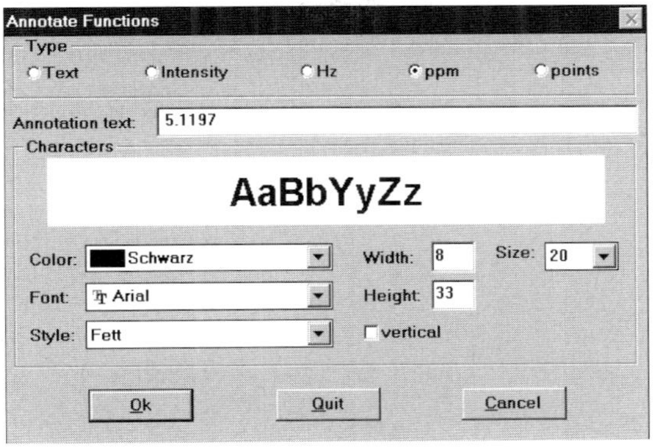

Fig. 4.27: Dialog box for annotation

4.9.3 Integration

Integration of 2D spectra is not as common as integration of 1D spectra. It is restricted to a few experiments, in particular NOESY and ROESY. The normalized cross peak integrals calculated from these type of spectra may be used quantitatively or at least semi-quantitatively to evaluate the transient NOEs and ROEs respectively and to subsequently obtain estimates of proton-proton internuclear distances. Usually a whole series of NOE and ROE experiments are performed, increasing the mixing time for the NOE (ROE) build-up from experiment to experiment. Integration of the corresponding cross peaks in this series of spectra allows the NOE (ROE) build-up to be followed as a function of the mixing time and so to establish the corresponding build-up curves. This data can then be used to extract structural information relating to internuclear distances.

Since integration is less common with 2D spectra only a short overview of the features in 2D WIN-NMR will be given. Activating the **Manual Integration** option in the **Analysis** pull-down menu changes the buttons menu on the lower part of the button panel (Fig. 4.28).

Define/Resize/ Shift Allows you to define a new integration region and to shift/resize it with the aid of the mouse. Several integral regions may be defined and are numbered accordingly (Fig. 4.29). To define integral regions of the same size, a mouse option for copying integration regions is available.

Delete/Delete All, Load Int./Save Int., Normalise The commmands allow you to delete, save/load and to normalize (e.g. with respect to a diagonal peak) integral regions of 2D spectra.

Report A full report of the defined integrals can be listed in a report dialog box, including the size or width of the integral region (in data points), the start and end rows and columns (number and ppm), the integral value (absolute, normalised) and the calculation mode used.

Consult the corresponding **Help** routine for more detailed information and for how to use these integral options.

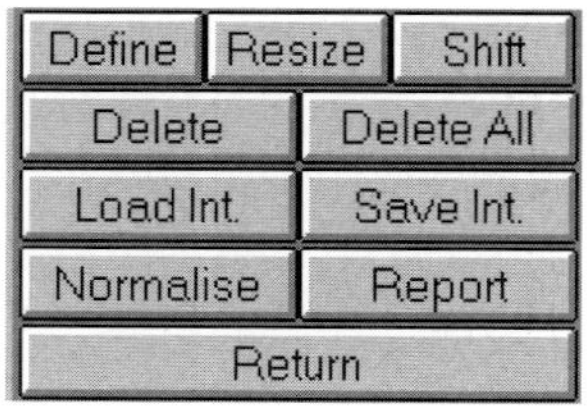

Fig. 4.28: **Manual Integration** buttons panel

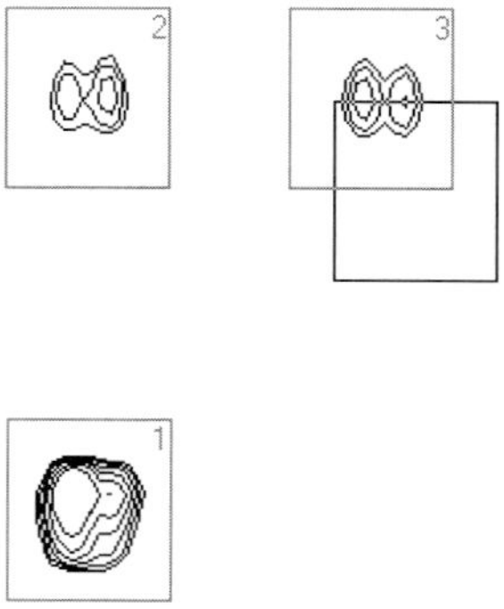

Fig. 4.29: Define integral regions

Define/Resize/ Shift Allows you to define a new integration region and to shift/resize it with the aid of the mouse. Several integral regions may be defined and are numbered accordingly (Fig. 4.29). To define integral regions of the same size, a mouse option for copying integration regions is available.

Delete/Delete All, Load Int./Save Int., Normalise The commmands allow you to delete, save/load and to normalize (e.g. with respect to a diagonal peak) integral regions of 2D spectra.

Report A full report of the defined integrals can be listed in a report dialog box, including the size or width of the integral region (in data points), the start and end rows and columns (number and ppm), the integral value (absolute, normalised) and the calculation mode used.

Check it in 2D WIN-NMR:

Load the ROESY spectrum of peracetylated glucose D:\NMRDATA\GLUCOSE\2D\HH\GHHRO\001999.RR and try to evaluate the volume integrals for the diagonal peak of the anomeric proton and its ROE peaks along F1 and F2. Select the **Manual Integration** option in the **Analysis** pull-down menu. Start with the diagonal peak, click the **Define** button, define an appropriate integral region and save it by pressing the right mouse button inside the rectangle (its borders change to red). Position the mouse cursor within this region, click the left mouse button which automatically generates a second integral region of the same size and drag it with the left mouse button pressed to the next position, i.e. to one of the ROE peaks of the anomeric proton. Proceed the same way to define integral regions of identical size for all the remaining ROE cross peaks of the anomeric proton in F1 and F2

Click on the **Save Int.** button and save these integral regions. Try out the **Resize**, **Shift** region, **Delete** region and **Delete All** buttons. Load the saved integral regions again (**Load Int.**) and normalize the integrals (**Normalise**) with respect to the (negative) diagonal peak (–100). Set **New reference integral**, **New mode** and **New reference value** in the **Normalise** dialog box to 1, – and 100 respectively. Exit this dialog box and click on the **Report** button to inspect the integral listing.

Compare the ROE data with the corresponding data obtained from the 1D ROE spectra.

4.10 Plotting 2D Spectra

As already outlined for 1D spectra, plotting of NMR data is by no means a task of secondary importance. The choice of suitable expansions, the completeness of additional data, including the relevant acquisition, processing and plot parameters, an appropriate title and a structural formula if available facilitate the subsequent spectral analysis.

This section gives an overview of the most important options for plotting spectra available in the **Output** pull-down menu of 2D WIN-NMR (Fig. 4.30). These options affect the output of the 2D spectrum, its projection spectra in F1 and F2, the title text and the parameter lists. The 2D spectrum and corresponding projection spectra, either calculated from the 2D data itself or separately measured as a standard 1D spectrum can be displayed on the screen in the 2D WIN-NMR window or sent directly to a hardcopy device.

An additional feature, the *Preview* window, accessible with the **Preview** button from within the **Page Setup...** sub-menu in the **Output** pull-down menu, displays an exact copy of your plot (WYSIWYG). It is recommended that you exploit this useful option and inspect the layout in this *Preview* window prior to plotting your 2D NMR spectra in order to avoid a waste of paper and time.

To prepare a plot layout the corresponding spectrum must first be displayed on the screen and a few display options (section 4.8), including the spectral limits, the number of levels and their values used in the display, the definition of projection spectra and annotation, must be set at this stage, and final processing (section 4.9) must be performed. Additional features dedicated to setting up or modifying the layout, to initialize plotting, or to use the WINDOWS clipboard for exporting your spectral layout to other application programs or for importing the contents of the clipboard generated with other application programs, are available with the **Output** pull-down menu of 2D WIN-NMR and will be discussed below.

Compared to 1D WIN-NMR the scope of the output options available with 2D WIN-NMR is less comprehensive. The inclusion of graphics elements (e.g. structural formulae), additional text files (e.g. pulse programs) and the interactive drawing of lines and rectangles is limited or not possible with 2D WIN-NMR. However the more powerful layout capabilities of 1D WIN-NMR may also be exploited by 2D data sets (section 4.10.5).

Fig. 4.30: **Output** pull-down menu

Output features available within the **Output** pull-down menu of 2D WIN-NMR:

4.10.1 Layout

This allows you to store your layout (**Save layout...**) and to reload it (**Load layout...**) for later use in order to get standardised plots of your 2D spectra. With these commands the kind of windows (spectrum, title, parameter), their position and size and additional features (axis, units, projections, parameters, ...) are stored and loaded respectively. After loading a layout, these layout parameters may be inspected and if necessary modified in the **Page Setup** dialog box opened with the **Page Setup...** command (section 4.10.2) before a data set is loaded and plotted.
Note: Defaults are set for the fonts, their style and color which are used for the title and the parameters. They cannot be modified by the user.

4.10.2 Page Setup...

Page Setup... opens a dialog box (Fig. 4.31) which allows you to inspect the current layout and print options.

The **Sizes** field contains a graphical representation of the currently selected page size and its orientation, as defined by the **Printer Setup...** menu (see 4.10.3). The plot elements are arranged on the page either with the sizes defined in the adjacent **Text** and **Trace** fields, or automatically (**Auto Arrange** button), which optimizes the size to fill the hardcopy sheet.

In general, the page layout consists of these three individual windows, the **Text** window for the title, the **Trace** window for the 2D matrix, with space left for projections, and the **Parameter** window.

Use the **Help** routine of 2D WIN-NMR for detailed information about the various edit fields and buttons available in the **Page Setup...** dialog box if necessary.

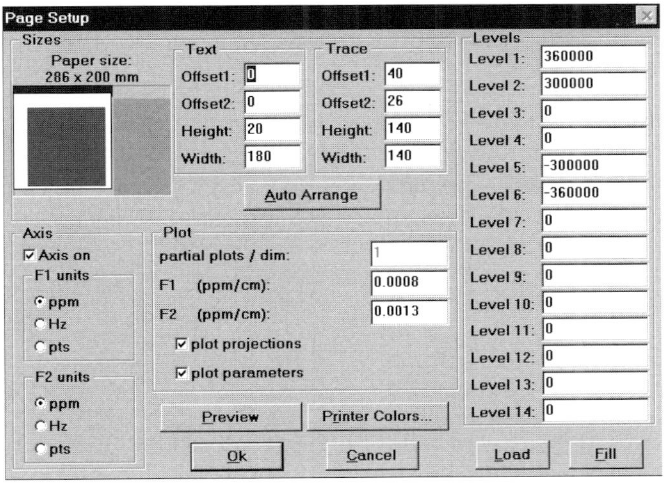

Fig. 4.31: The **Page Setup** dialog box.

4.10.3 Print... , Print all, Printer Setup...

To start the printout operation the **Print...** option in the **Output** pull-down menu is used. If there is more than one window on the screen choosing the **Print...** option – and using the mouse – allows you to specify the window which should be plotted. If all windows should be plotted, use the **Print all** option. With **Printer Setup...** a variety of output device specific parameters defining the resolution, the paper size and other printer variables, may be set.

Attention: If the plotted spectrum does not correspond to the preview display, e.g. only part of the spectrum is plotted, and an error message appears on your output device, readjust the set up of the 2D spectrum. It may be that the memory size of your output device is not large enough with respect to the required printing, that the option for a full page plot is disabled or that a configuration parameter for the output device has not been set correctly.

4.10.4 Copy, Copy all, Paste

With the **Copy** command a copy of the screen contents is stored in the MS-WINDOWS clipboard. If you have more than one window on the screen, the window to be copied may be specified using the mouse. If all windows should be copied into the clipboard use the **Copy all** command. The **Copy** options are used to transfer 2D layouts into the *Preview* window of 1D WIN-NMR (see 4.10.5).

With the **Paste** command any contents (ASCII format, bitmap, metafile) from the clipboard may be copied into one of the 2D WIN-NMR windows.

4.10.5 2D Layout with 1D WIN-NMR

As an alternative to using the 2D WIN-NMR layout, the more powerful and versatile layout features of 1D WIN-NMR may be exploited. The preliminary layout is prepared with 2D WIN-NMR and transferred via the clipboard (see section 4.10.4) into 1D WIN-NMR using the **Copy** and **Paste** options respectively. The final layout (Fig. 4.32) is then completed with additional elements such as text files containing the pulse program, lists with extended parameters, graphics files containing the structural formula, and interactively set up graphics (lines, rectangles).

Follow the instructions given in one of the subsequent *Check its* or refer to the section 4.7.3, where the use of the 1D WIN-NMR *Preview* window is described, for how to do this.

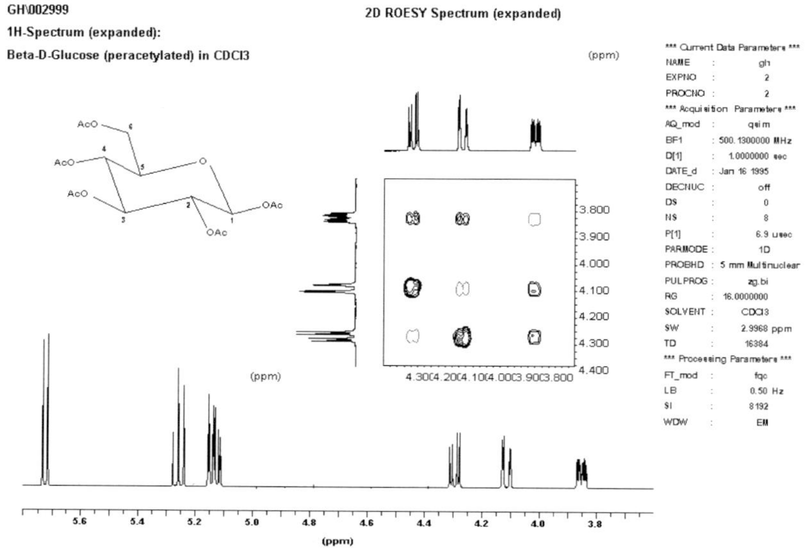

Fig. 4.32: Example of a layout with a 2D spectrum established with 2D WIN-NMR and 1D WIN-NMR.

4.10.6 History...

All changes to the processed data set are registered in a special history file. This file can be inspected and, if required, changed by means of the **History...** menu item. On activating this command a history dialog box appears on the screen with options for saving and printing.

Check it in 2D WIN-NMR: H/H-COSY (magnitude mode)

Load the gradient assisted 2D magnitude mode $^1H/^1H$ COSY spectrum of peracetylated glucose D:\NMRDATA\GLUCOSE\2D\HH\GHHCO\002999.RR. Check and if necessary correct its calibration with respect to the calibrated, separately measured projection spectrum D:\NMRDATA\GLUCOSE\2D\1DREF\GH\001999.1R). Set up a display with projections in both F1 and F2, adjust the relative sizes of the 2D spectrum and the projections and activate the grid to facilitate subsequent spectral analysis. Choose six levels and adjust their values for the best presentation using the appropriate buttons in the button panel. Use the automatic procedure for adjusting these levels as well and compare the results.

Choose the **Page Setup...** option in the **Output** pull-down menu and select the plotting parameters and projections to be included in the plot. Adjust the sizes of all windows (title, parameter, spectrum) to the size of the paper sheet. Set up a title and select the parameters to be plotted. Load the defined levels with the **Load** button, choose different colors for the levels if you have a multicolor output device. Click the **Preview** button to inspect the final layout in the *Preview* window.

From the **Display** pull-down menu choose the **Window** option and create an additional window within the 2D plot to include the structural formula of the glucose sample. Start a parallel 1D WIN-NMR session and load the structure file D:\NMRDATA\GLUCOSE\GLUCOSE.WMF) into the 1D WIN-NMR *Preview* window. Copy it into the clipboard with the **Copy** command available in the **Edit** pull-down menu of 1D WIN-NMR. Return to 2D WIN-NMR, activate the additional window and choose the **Paste** option in the **Output** pull-down menu. Move the mouse cursor into the additional window and initialize the transfer from the clipboard by clicking the left mouse button. Resize this window if necessary.

Set up your output device, and start plotting the 2D spectrum with the **Print all** command.

Now load the basic 2D magnitude mode $^1H/^1H$ COSY spectrum D:\NMRDATA\GLUCOSE\2D\HH\GHHCO\001999.RR of peracetylated glucose, acquired without gradients, and produce in the same way a corresponding plot as above. Compare the two spectra with respect to their quality.

Check it in 2D WIN-NMR: H/H-COSY (magnitude mode)

Use again the 2D magnitude mode $^1H/^1H$ COSY spectrum D:\NMRDATA\GLUCOSE\2D\HH\GHHCO\001999.RR and switch to the *Stacked* display mode. Select a region of interest. Use the **Grid** button in the button panel and the mouse to move and tilt the "3D" spectrum according to your ideas. Set up your output device and plot the expanded COSY spectrum in stacked display mode.

Check it in 2D WIN-NMR: JRES

Load the $^{1}H/^{1}H$ 2D magnitude mode J-resolved spectrum of peracetylated glucose D:\NMRDATA\GLUCOSE\2D\HH\GHHJR\001999.RR. Check and if necessary correct its calibration in F2 with respect to the calibrated, separately measured projection spectrum D:\NMRDATA\GLUCOSE\2D\1DREF\GH\001999.1R). Set up a display with a F2 projection, adjust the relative sizes of the 2D spectrum and the projection and activate the grid to facilitate subsequent spectral analysis. Choose six levels and adjust their values for the best presentation using the appropriate buttons in the button panel. Eventually use the automatic procedure. Choose different colors for the levels if you have a multicolor output device. Add the structural formula if desired.

Choose the **Page Setup...** option and select the plotting parameters and projections to be included in the plot. Adjust the sizes of all windows (title, parameter, spectrum) to the size of the paper sheet. Set up a title and select the parameters to be plotted. Load the defined levels with the **Load** button and inspect the layout in the *Preview* window.

Set up your output device, and start plotting the 2D spectrum with the **Print all** command.

Check it in 2D WIN-NMR: C/H-COSY (magnitude mode)

Load the heteronuclear $^{1}H/^{13}C$ magnitude mode 2D COSY spectrum of peracetylated glucose D:\NMRDATA\GLUCOSE\2D\CH\GCHCO\001999.RR. Check and if necessary correct its calibration in F2 and F1 with respect to the calibrated, separately measured ^{1}H and ^{13}C projection spectra D:\NMRDATA\GLUCOSE\2D\1DREF\GH\001999.1R and D:\NMRDATA\GLUCOSE\2D\1DREF\GC\001999.1R). Set up a display with the two projection spectra, adjust the relative sizes of the 2D spectrum and the projections and activate the grid to fascilitate subsequent spectral analysis. Choose six levels and adjust their values for the best presentation using the appropriate buttons in the button panel or the "slice freeze" method (4.8.2). Eventually use the automatic procedure. Choose different colors for the levels if you have a multicolor output device.

Use the annotation feature to assign each cross peak with its shift values and add the structural formula if desired.

Enter the **Page Setup** dialog box, activate the option for plotting parameters and the projections and adjust the sizes of all windows (title, parameter, spectrum) to the size of the paper sheet. Set up a title and select the parameters to be plotted. Load the defined levels with the **Load** button and inspect the layout in the *Preview* window.

Set up your output device and start plotting the 2D spectrum with the **Print ...** command.

Check it in 2D WIN-NMR: C/H- long range- COSY (HMBC, magnit. mode)

Load the gradient assisted, inverse detected, heteronuclear $^{1}H/^{13}C$ magnitude mode 2D COSY spectrum of peracetylated glucose, dedicated to observing long-range coupling interactions D:\NMRDATA\GLUCOSE\2D\CH\GCHI COLR\001999.RR. Check and if necessary correct its calibration in F2 and F1 with respect to the calibrated, separately measured ^{1}H and ^{13}C projection spectra D:\NMRDATA\GLUCOSE\2D\1DREF\GH\001999.1R and D:\NMR DATA\GLUCOSE\2D\1DREF\GC\002999.1R). Set up a display with the two projection spectra, adjust the relative sizes of the 2D spectrum and the projections and activate the grid to fascilitate subsequent spectral analysis. Choose six levels and adjust their values for the best presentation using the appropriate buttons in the button panel or the "slice freeze" method (4.8.2). Eventually use the automatic procedure. Choose different colors for the levels if you have a multicolor output device.

Use the annotation feature to assign each cross peak with its shift values and add the structural formula if desired.

Enter the **Page Setup** dialog box, activate the option for plotting parameters and the projections and adjust the sizes of all windows (title, parameter, spectrum) to the size of the paper sheet. Set up a title and select the parameters to be plotted. Load the defined levels with the **Load** button and inspect the layout in the *Preview* window.

Set up your output device and start plotting the 2D spectrum with the **Print all** command.

Check it in 2D WIN-NMR: C/H- long range- COSY (HMBC, magnitude mode)

Repeat the same procedure to plot the other C/H-long-range COSY spectra:

002999	(**CH** / **C**O)
003999	(**C**H_3 / **C**O)
004999	(**CH** / **C**H)

Assign the ^{1}H and ^{13}C signals according to the information from previous spectra. Try to identify and separate residual $^{1}J_{CH}$ cross peaks (appearing as $^{1}J_{CH}$ doublets, centred around the corresponding ^{1}H chemical shifts) and to establish $^{1}H/^{13}C$ coupling connectivities over 2 and 3 bonds.

Check it in 2D WIN-NMR: H/H-TOCSY (phase sensitive)

Load the phase mode 2D TOCSY spectrum of peracetylated glucose D:\NMRDATA\GLUCOSE\2D\HH\GHHTO\001999.RR. Calibrate the spectrum according to the procedure outlined above for the COSY spectrum. For peracetylated glucose with only one coupled spin system, this experiment is not particularly helpful. In chapter 5, when processing the corresponding data measured for the peracetylated oligosaccharide the great advantage of the 2D TOCSY experiment will become apparent.

Set up the layout and plot the spectrum according to the guidelines given above and according to your ideas.

Check it in 2D WIN-NMR: C/H-COSY (HMQC, phase sensitive)

Load the gradient assisted, inverse detected 2D phase mode $^{1}H/^{13}C$ COSY (HMQC-) spectrum of GLUCOSE D:\NMRDATA\GLUCOSE\2D\CH\GCHICOMQ\001999.RR. Check and if necessary correct its calibration in both dimensions with respect to the calibrated, separately measured ^{1}H- and ^{13}C projection spectra D:\NMRDATA\GLUCOSE\2D\1DREF\GH\001999.1R and D:\NMRDATA\ GLUCOSE\ 2D\1DREF\GC\001001.1R. Set up a display with the spectral limits from 94 to 56 ppm in F1 and from 5.9 to 3.4 ppm in F2. Adjust 4 contour levels for the best presentation and activate the grid. Choose and activate the projections for F1 and F2 and adjust the relative sizes of the 2D spectrum and the projections. Open the **Page Setup** dialog box, activate the parameter window for the layout and fit the size of your 2D spectrum and its projections to the paper size. Select the most important F1 and and F2 acquisition, processing and plot parameters and define a title. Use the **Load** button to include all four contour levels to be plotted. Inspect the final layout in the *Preview* window and – after setting up the output device – start plotting.

Use the **Copy all** function in the **Output** pull-down menu to copy this layout into the clipboard. This copy will be used in the next *Check it*. Note that with **Copy** instead of **Copy all** only the active window, e.g. the contour plot window, but not the other windows will be copied into the clipboard.

Assign the ^{13}C signals with the ^{1}H assignments known from previous experiments. Compare the spectrum with the spectrum of the basic, ^{13}C-detected CH-COSY spectrum.

Check it in 1D WIN-NMR:

Start 1D WIN-NMR, enter the *Preview* window and use the **Frame** option to open a frame for the 2D layout copied into the clipboard in the previous *Check it*. With the cursor positioned within the frame double click on the left mouse button to open the dialog box. Click on **Paste** and then **OK** to import the 2D layout. Double click the left mouse button and in the **Metafile Options** dialog box adjust **x-Factor** and **y-Factor** to position the 2D layout correctly within the frame. Open additional frames to accommodate the entire 1D ^{13}C spectrum D:\NMRDATA\GLUCOSE\1D\C\GC\001999.1R) and the structural formula (Fig. 4.33). Arrange and resize these frames for the best representation. Use additional graphical elements (Lines, Rectangle) available within the *Preview* window of 1D WIN-NMR for assignment purposes. Set up your output device and plot the layout.

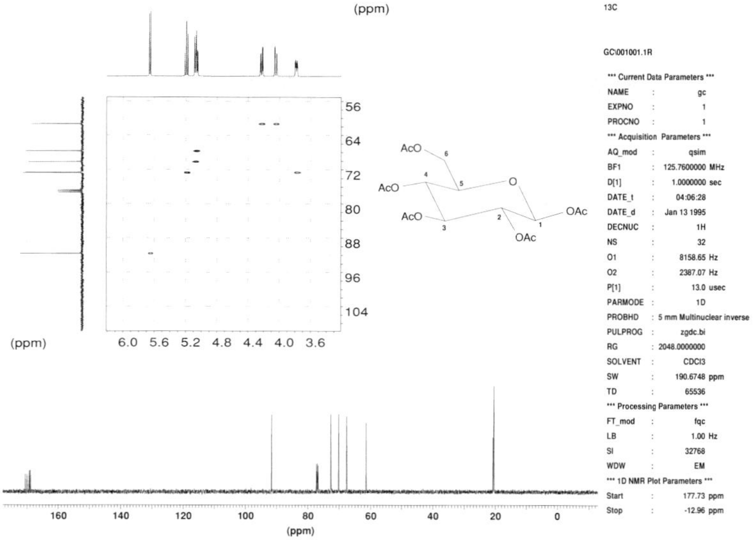

Fig. 4.33: Layout obtained with 1D WIN-NMR, including a 2D spectrum.

Check it in 2D WIN-NMR/1D WIN-NMR:

Repeat the procedure outlined in the last two *Check its* to plot the second HMQC spectrum D:\NMRDATA\GLUCOSE\2D\CH\GCHICOMQ\ 002999.RR, showing the expansion in the region of the methyl carbon and proton signals.

Check it in 2D WIN-NMR: C/H-COSY (HSQC, phase sensitive)

Load the spectra of the gradient-assisted, inverse detected, 2D CH-HSQC experiments acquired without and with the echo-antiecho technique, D:\NMRDATA\GLUCOSE\2D\CH\GCHICOSQ\001999.RR and D:\NMRDATA\ GLUCOSE\2D\CH\GCHICOSQ\002999.RR respectively. Process them and set up layouts as for the basic HMQC spectrum. Compare the spectra obtained for these three experiments with respect to their sensitivities. Use the same rows or columns to quantify the sensitivity gain obtained with the echo-antiecho technique. All three spectra were acquired with the same measuring time using the same experimental parameters.

Check it in 2D WIN-NMR: H/H-COSY (DQ-filtered, phase sensitive)

Load the gradient assisted double quantum filtered phase sensitive COSY spectrum of peracetylated glucose D:\NMRDATA\GLUCOSE\2D\HH\ GHHCODF\002999.RR. Process and set up the layout for this spectrum as above. Select the option for displaying positive and negative contour levels and adjust their values. Use two colors (black/red) for displaying (and eventually plotting) positive and negative levels respectively.

Open an additional window within the 2D spectrum and include an expansion of a cross peak. Use the *Stacked* display mode to display this cross peak within this additional window, whereas the *Contour* display mode remains still active for the main window. Include the structural formula of peracetylated glucose if desired and plot the spectrum according to your preferred layout ideas.

Try again to establish the coupling network and compare the results with the results obtained with the magnitude mode COSY experiment. Plot individual rows and columns and try to differentiate between active and passive couplings and to get numerical values for the corresponding coupling constants. Store these rows and columns for later use in the next *Check it.* Using the **Dual Display** option of 1D WIN-NMR compare one of the rows of this phase-sensitive spectrum with the corresponding row in the basic magnitude 2D COSY spectrum.

Check it in 2D WIN-NMR: H/H-COSY (DQ-filtered, phase sensitive)

Load the basic double quantum filtered phase sensitive COSY spectrum of peracetylated glucose, acquired without gradients D:\NMRDATA\GLUCOSE\2D\HH\GHHCODF\001999.RR. Choose the corresponding rows and columns as in the previous *Check it* and compare them. Notice again the higher spectral quality obtained with the use of gradients.

Check it in 2D WIN NMR: H/H-ROESY (phase sensitive)

Load the phase sensitive 2D ROESY spectrum of peracetylated glucose D:\NMRDATA\GLUCOSE\2D\HH\GHHRO\001999.RR and calibrate it according to the procedure outlined above for the COSY spectra. Use two colors (black/red) for displaying positive (cross peaks) and negative (diagonal peaks) levels respectively. Define the number of levels and adjust their values. Decrease the contour level values for the negative diagonal peaks by at least a factor of 10. Include the structural formula if desired and plot the spectrum according to your preferred layout ideas.

Assign the signals using the COSY information, and try to establish the dipolar coupling network, i.e. to find spatial proximity between protons. Compare the results with the results extracted from the spectra of the corresponding 1D ROE experiments.

Check it in 2D WIN-NMR:

Load the 2D NOESY spectrum of peracetylated glucose D:\NMRDATA\GLUCOSE\2D\HH\GHHNO\001999.RR. Process it and set up a layout according to the procedure outlined above for the 2D ROESY spectrum. Plot the spectrum according to your preferred layout ideas. Compare the results with the results obtained with the 2D ROESY and the 1D NOE experiments.

5 How to Process 1D and 2D NMR Data

5.1 Introduction

When submitting a sample for NMR analysis (measured either automatically or manually) you usually have access to the plotted spectrum and sometimes the raw spectral data (FID). Although it is not necessary to process the data yourself, there are a number of important reasons why you should become familiar with the basic principles and rules of NMR data processing. Generally data processing is applied:

- to make the NMR data readable and to prepare it for subsequent analysis.
- to calibrate and reference the NMR data.
- to manipulate the NMR data.
- to prepare the layout for subsequent plots of the NMR data.

Probably the simplest and most obvious reason for processing is to make the NMR data readable, allowing you to extract information from the spectrum to solve your structural problems. The FID, i.e. the time domain signal s(t), is transformed into a spectrum i.e. the frequency domain signal S(F), using a Fourier transformation (FT). Your NMR spectrum can then be compared with reference data. Additional information can be extracted from the spectrum after calibration, peak picking and, if necessary, integration of the spectral regions. These items have been discussed already in chapter 4.

The data may be manipulated in a number of ways to assist the subsequent spectral analysis. The two most important options are the enhancement of the signal-to-noise ratio in noisy spectra (with subsequent reduction in spectral resolution) and the converse, the improvement of spectral resolution (with subsequent loss in the signal-to-noise ratio). Experiments such as decoupling and NOE difference require special arithmetic operations such as the addition or subtraction of different data files to high-light the subtle changes in your spectrum. Furthermore, unwanted signals, originating from spectrometer instabilities, or simply caused by incorrectly set experimental parameters, can be suppressed or even eliminated with suitable data processing.

Finally, the preparation of the plot layout is determined by the problem under investigation, the type of experiment performed and your own preferences. Suitable expansions for both 1D and 2D spectra not only simplify the analysis, but make signal assignments and the evaluation of small coupling constants more reliable.

Data manipulations can be performed before and after the Fourier transformation. Normally you would start with a series of manipulations on the time-domain signal, apply the Fourier transformation before performing several final processing steps in the frequency domain. The procedure for processing 1D and 2D data is obviously different; 1D data requires only one Fourier transformation while 2D data requires two Fourier transformations. In addition, processing options, only meaningful for 2D data, exist. However in both cases the basic principles and guidelines for improving the signal-to-noise ratios, enhancing spectral resolution, referencing spectra and preparing layouts for subsequent plots are the same.

The main steps for transforming raw 1D and 2D data into the corresponding spectra are shown in Figs. 5.1 and 5.2 respectively.

In general there is one important point regarding data processing which must be emphasized. For the correct processing of NMR data you should be familiar with the general rules of data processing and should understand the basics of the corresponding experiment. Failure to understand these principles can effect the quality and the reliability of the NMR parameters extracted in the subsequent spectral analysis and, in the worst case, can lead to the wrong conclusion being drawn regarding the final structure.

In this chapter the various processing options available with 1D WIN-NMR and 2D WIN-NMR are presented. In the first part the basic steps required to produce a spectrum, i.e. Fourier transformation and phasing the spectrum, are discussed. The second part is devoted to more specialised processing steps, usually applied to enhance the qualitiy of your NMR data and to tailor your final spectrum to the requirements of the subsequent data analysis. The many processing options are applied to the raw data of peracetylated β-D-glucose and their effects may be studied following the instructions given in the corresponding *Check its*. In addition to following these instructions you are strongly advised to experiment with these processing tools and to try out the other processing parameters as well. In chapter 6 you can test your skills in data processing using the raw NMR data for a peracetylated oligosaccharide of unknown structure. This data has been acquired using the same series of experiments as for the peracetylated glucose and should encourage you to apply what you have learnt on a more demanding sample.

1D DATA

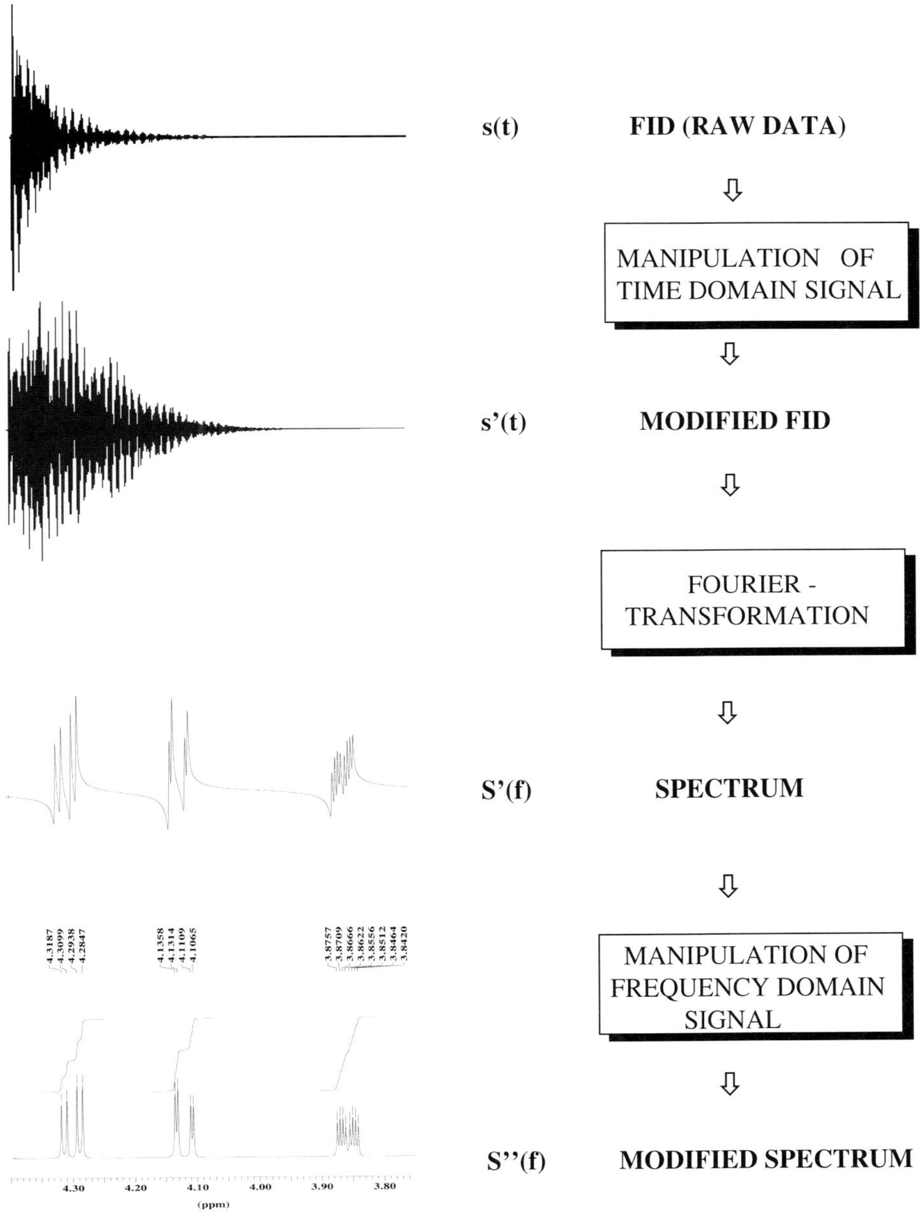

Fig. 5.1: Data processing steps with 1D NMR raw data.

2D DATA

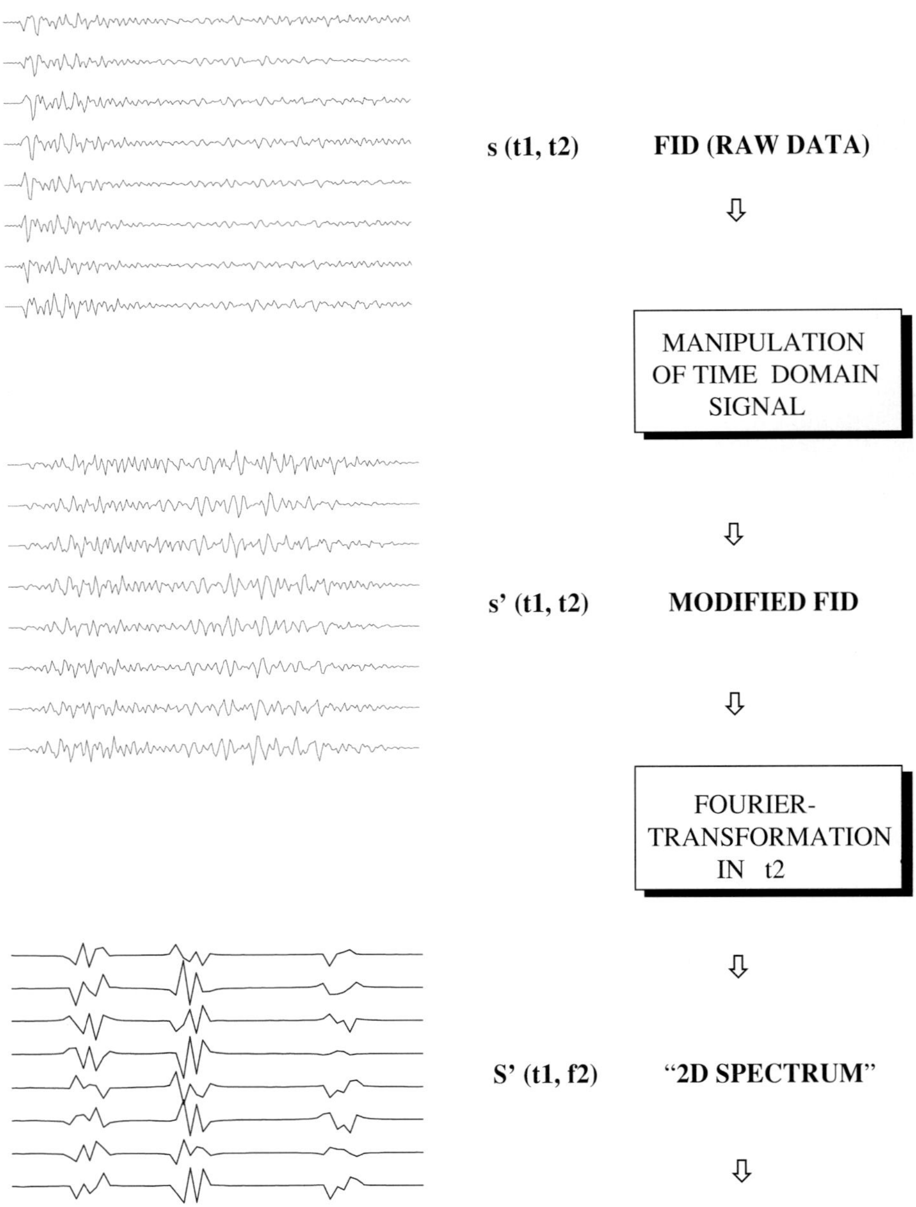

Fig. 5.2: Data processing steps with 2D NMR raw data.

2D DATA (continued)

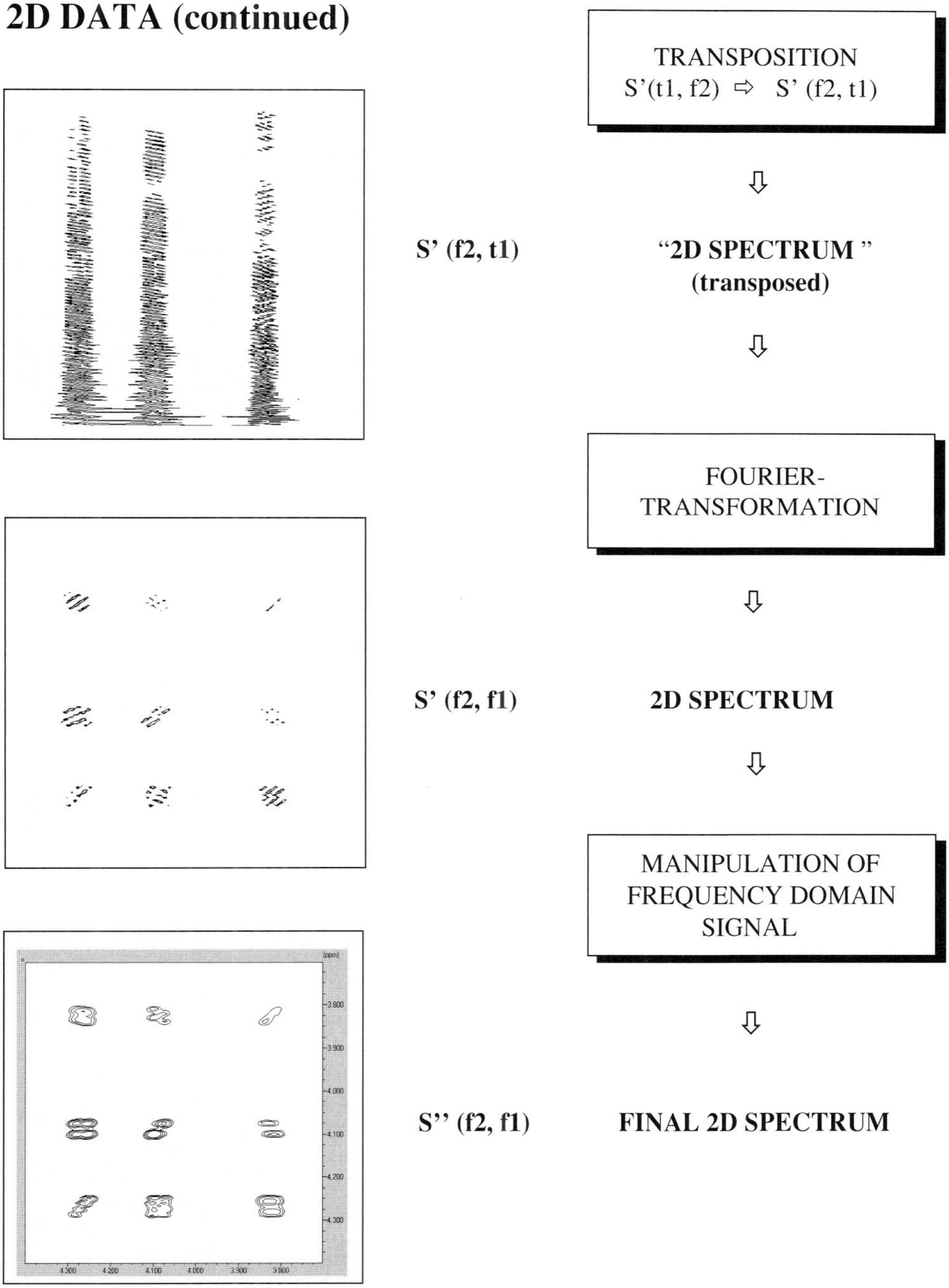

Fig. 5.2 (continued): Data processing steps with 2D NMR raw data.

5.2 Basic Processing

The basic processing of 1D and 2D data requires obligatory processing steps for transforming the raw data (FID) into a "readable" spectrum, i.e. Fourier transformation and phase correction to produce a spectrum with absorptive lineshapes. Finally, a few additional steps (calibration, peak picking, integration) as discussed in chapter 4 are required before the spectrum is eventually plotted.

5.2.1 The Parameters TD and SI

The response of any 1D or 2D NMR experiment is detected by a receiver coil in the probehead. This signal is amplified in the receiver, digitised in an analog-to-digital converter (ADC) and stored as a series or a whole matrix of discrete data points in the memory of your spectrometer computer. The data is not continuous and discrete processing has to be used to modify and to convert this digitised raw data into a spectrum. With 1D experiments, using quadrature detection (see volume 2 of this series), an acquisition time **AQ** and a spectrum width **SW**, a series of **TD** complex data points are acquired. There are **TD/2** real data points and **TD/2** imaginary data points which are stored separately.

$$\mathbf{TD = 2\,SW \cdot AQ}$$

The length of **AQ** determines the spectral resolution observed in a spectrum. The longer **AQ** the higher the spectral resolution.

Prior to Fourier transformation the actual number of measured data points and the total number of data points to be transformed may be defined. With 1D WIN-NMR the following definitions are valid and may be inspected for the actual data file by selecting the **Zero Filling...** option in the **Process** pull-down menu.

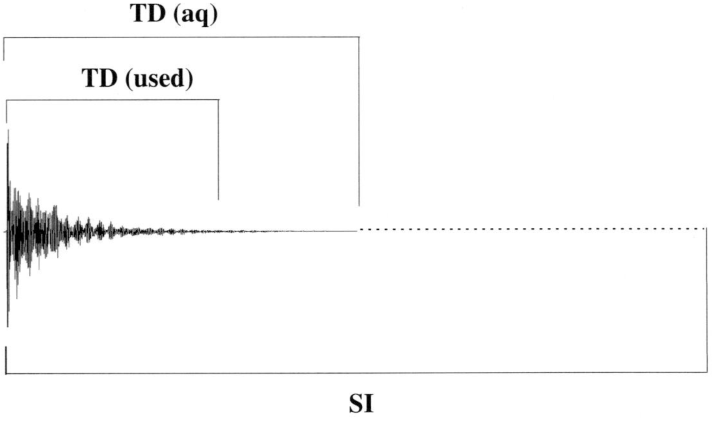

TD(aq) Is the total number of measured data points in the FID i.e. **TD(aq)/2** real and **TD(aq)/2** imaginary data points.

TD(used) Is the effective number of measured data points used for the calculation. When **TD(used)** < **TD(aq)** the residual data points are set to zero. **TD(used)** does not have to be a power of 2.

SI(r+i) Is the sum of the real and imaginary data points that the resulting FID and the spectrum will consist of (i.e. **SI/2** real and **SI/2** imaginary data points). **SI(r+i)** determines the digital resolution (Hz/data point) of a spectrum and must be a power of 2 e.g. 4096 or 4k, and may be larger than **TD(aq)** or **TD(used)** as described in section 5.3.3 ("Zero Filling").

With 2D experiments the situation is a little more complicated as the size of the overall digitised matrix depends on the number of time increments in t1 as well as parameters specific to the 2D acquisition mode. Nevertheless, a digitised matrix of **TD(2)** x **TD(1)** complex data points is acquired and stored. Similar to 1D the effective number o measured data points used for calculation **TD(used)** and the total number of data points **SI** to be transformed in t2 and t1 may be defined prior to Fourier transformation. These parameters may be inspected and defined in the **General parameter setup** dialog box accessible via the **Process** pull-down menu. With 2D WIN-NMR the definitions for **TD(2)** and **TD(1)** are the same as for **TD** with 1D WIN-NMR. However, unlike 1D WIN-NMR, with 2D WIN-NMR **SI(2)** and **SI(1)** define the number of pairs of complex data points, instead of the sum of the number of real and imaginary data points. Therefore the 2D FT command (see below) transforms the acquired data of the current data set into a spectrum consisting of **SI** data points in both the real and the imaginary part.

5.2.2 Fourier Transformation of 1D Data

The central step of any NMR data processing is the Fourier transformation (FT) which transforms the time domain signal s(t) – the raw data – of a 1D experiment into a frequency domain signal S(f) – the spectrum:

$$\mathbf{S\,(f) = \int s\,(t) \cdot e^{-i \cdot f \cdot t}\, dt} \qquad \text{complex Fourier transformation}$$

with the time domain signal (quadrature detection) of a single resonance line at the resonance frequency F and characterized by the "relaxation" time T:

$$\mathbf{s\,(t) = s^0 \cdot e^{i \cdot F \cdot t} \cdot e^{-1/T \cdot t}}$$

With digitized data a "discrete" Fourier transformation is performed:

$$\mathbf{S\,(f) = 1/N \cdot \sum_{k=0}^{N-1} s\,(t) \cdot e^{-i \cdot f \cdot k / N}}$$

The time domain signal, when measured in quadrature detection mode (see volume 2 in this series) in order to discriminate between positive and negative frequencies (+F, –F), is complex and yields after a complex Fourier transformation a real and an imaginary part, R(f) and I(f) respectively. With an ideal spectrometer the real part corresponds to an absorptive and the imaginary part to a dispersive Lorentzian lineshape:

$$\mathbf{R\,(f)} = \mathbf{S^0}\,\frac{\mathbf{1/T}}{\mathbf{1/T^2} + \mathbf{(F-f)^2}} = \mathbf{A\,(f)} \qquad \text{absorptive Lorentzian lineshape}$$

$$\mathbf{I\,(f)} = \mathbf{S^0}\,\frac{\mathbf{(F-f)}}{\mathbf{1/T^2} + \mathbf{(F-f)^2}} = \mathbf{D\,(f)} \qquad \text{dispersive Lorentzian lineshape}$$

Fourier transformation in 1D WIN-NMR is accomplished by choosing either the **FT** command in the **Process** pull-down menu (Fig. 5.3) or by simply clicking the **FT!** button in the button panel.
Both Fourier transform commands perform a Fast Fourier Transform (FFT) on the FID. If a baseline correction has not yet been performed on the raw data, a message box will appear which provides the option for performing a baseline correction (see section 5.3.3) on the time domain data (FID) prior to Fourier transformation.

Clicking with the right mouse button on the **FT!** button opens a dialog box for activating and performing a 5th order phase correction, together with the FT. This automatically corrects non-linear phase distorsion in the spectrum, introduced by electronic filters. With the available data this correction is not necessary and its application produces no effects in the final spectrum.

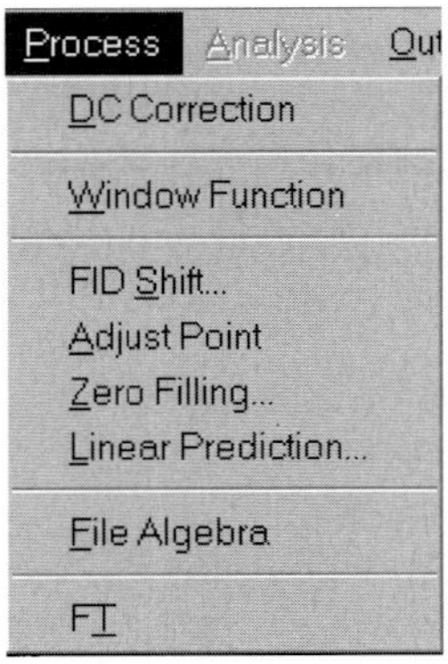

Fig. 5.3: **Process** pull-down menu.

5.2.3 Phasing of 1D Spectra

In reality the individual lines obtained after the Fourier transformation are composed of both absorptive A(f) and dispersive D(f) components. This "non-ideality" arises because of a phase shift between the phase of the radiofrequency pulses and the phase of the receiver, **PHC0**, and because signal detection is not started immediately after the excitation pulse but after a short delay period Δ. Whereas the effect of the former is the same for all lines in a spectrum and can be corrected by a zero-order phase correction **PHC0**, the latter depends linearly on the line frequency and can be compensated for by a first-order phase correction **PHC1**. Both corrections use the separately stored real and imaginary parts of the spectrum to recalculate a pure absorptive spectrum.

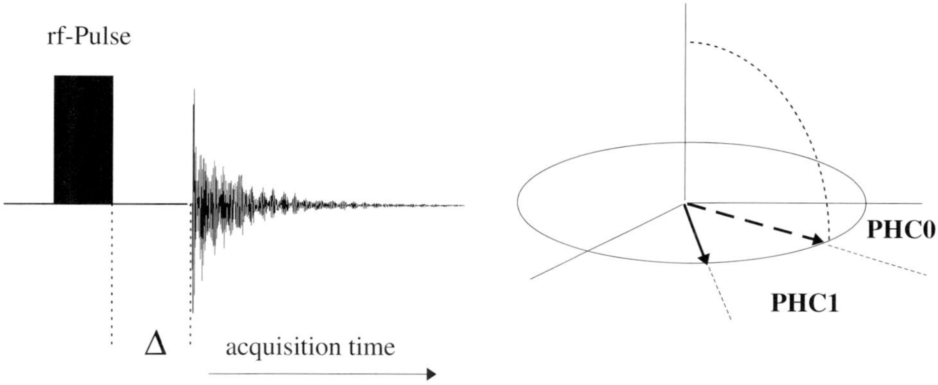

$$s(t) = s^0 \cdot e^{i[F \cdot t + (PHC0 + PHC1 \cdot \Delta)]} \cdot e^{-1/T \cdot t}$$

With 1D WIN-NMR several methods for phase adjustments of 1D spectra exist:

- With the **Autom. Phase Correction** option selected from the **Process** pull-down menu (Fig. 5.4) a fully automatic phase correction can be performed.
- If for a series of spectra the phase correction values **PHC0** and **PHC1** are the same and if these values have been determined, either automatically or manually (see below) for one spectrum, then the phasing of the remaining spectra may simply be accomplished with the **Phase Corr.!** button in the button panel (Fig. 4.2). Clicking with the right mouse button on this button, opens a dialog box to inspect the phase mode and to inspect/adjust the correction values **PHC0** and **PHC1** before the correction is initialized with the **Execute** button in this dialog box.
- With the command **Phase Correction** in the **Process** pull-down menu the button panel is switched into *Phase* mode (Fig. 5.4) and a semi-automatic or fully manual phase correction may be performed.

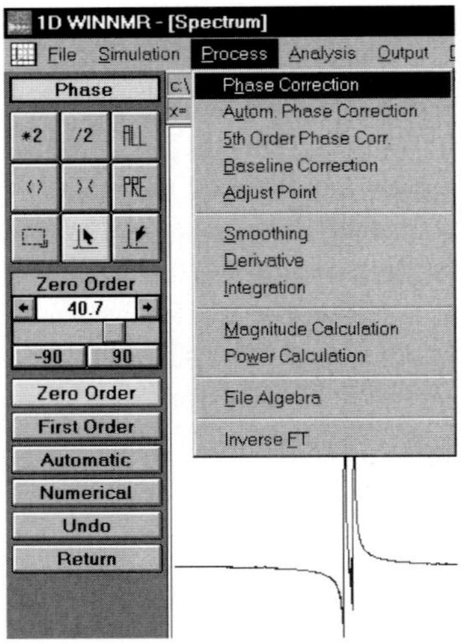

Fig. 5.4: Button panel for phase correction (left) opened by choosing the **Phase Correction** option in the **Process** pull-down menu (right).

The zero and first-order phase correction parameters may be modified manually either by entering numerical values in a dialog box (**Numerical** button), or in an interactive way, using the **Zero Order** and **First Order** buttons and their corresponding slider box. With the **Automatic** button a fast and rough phase correction is performed which speeds up the subsequent manual fine adjustments. This "automatic" phase correction is not the same as the fully automatic (and more time consuming) phase correction mentioned above.

The phase correction parameters **PHC0** and **PHC1** are always based on the initial spectrum obtained directly after a Fourier transformation operation. Phasing is performed with respect to a reference point which is marked on the spectrum by the spectrum cursor. When the *Phase* mode is first entered, the program automatically places the spectrum cursor on the tallest point in the spectrum, but you are free to move this spectrum cursor anywhere in the spectrum using the **Maximum Cursor** or the **Perpendicular Cursor** option.

Check it in 1D WIN-NMR:

Load the raw data of the 1D ^{1}H experiment measured for peracetylated glucose D:\NMRDATA\GLUCOSE\1D\H\GH\002001.FID and perform a Fourier transformation. Use either the **FT!** button in the button panel or from the **Process** pull-down menu choose the **FT** option. In the **DC Correction** dialog box click on the **No** button. Note that the calculated spectrum is incorrectly phased. Use the dual display option to compare this spectrum, showing the ring protons, with the correctly phased spectrum D:\NMRDATA\GLUCOSE\1D\H\GH\002999.1R. Exit the dual display and from the **Process** pull-down menu choose the **Phase Correction** option. The cursor will be positioned on one of the doublet lines of the anomeric proton. Click on the **Zero Order** button in the button panel and adjust the zero-order phase using the scroll bar. Select and expand one of the signals at the other end of the spectrum. Click on the **First Order** button and adjust the first-order phase correction in the same way. If you are satisfied with the result, note the numerical values for **PHC0** and **PHC1**, click the **Return** button and store the spectrum (...\002001.1R). Load, and Fourier transform the same raw data. From the **Process** pull-down menu choose the **Autom. Phase Correction** option and perform an automatic phase correction. Compare the automatically calculated correction factors **PHC0** and **PHC1** Compare the spectrum with the results obtained for the manual phase adjustment in the dual display.

Load the raw data of the 1D ^{1}H experiment measured for peracetylated glucose D:\NMRDATA\GLUCOSE\1D\H\GH\001001.FID and perform a Fourier transformation. Inspect the spectrum, showing the entire proton range, and adjthe phases to get pure absorption lines throughout. Do noit forget to store hthe spectrum (...\001001.1R).

Check it in 1D WIN-NMR:

Load and Fourier transform the raw data of the 1D ^{13}C DEPT experiment measured for peracetylated glucose D:\NMRDATA\GLUCOSE\1D\C\GCDP\001001.FID. Following the procedure outlined above using the **Phase Correction** option in the **Process** pull-down menu process the data to obtain a phased spectrum. Load the second and third data set and process it in the same way as above (same PHC0, PHC1), but use the buttons **FT!** and **Phasing!** for rapid processing. Click with the right mouse button on these buttons to first inspect the corresponding parameters before you initialize the individual processing steps with the left mouse button. Store all the spectra (...\001001.1R - (...\003001.1R).

5.2.4 Fourier Transformation of 2D Data

The time domain signal of a single resonance line as obtained with a 2D NMR experiment may be described as

$$\mathbf{s\,(t1, t2) = s^0 \cdot e^{\,i \cdot F(1) \cdot t1} \cdot e^{\,-1/T(1) \cdot t1} \cdot e^{\,i \cdot F(2) \cdot t2} \cdot e^{\,-1/T(2) \cdot t2}}$$

F(1), F(2) denote the resonance frequencies of the single resonance line in the two dimensions F1 and F2 respectively and T(1), T(2) characterize the relaxation behaviour of this line in the time domains t1 and t2 respectively.

Compared to the Fourier transformation of 1D data, where usually a complex FT is applied to the time domain signal, several procedures are used to perform a 2D Fourier transformation. The aims followed with 2D data acquisition and 2D Fourier transformation are twofold: quadrature detection in both dimensions and spectra with pure absorption lines in both dimensions. With quadrature detection in both dimensions and the transmitter frequency centred in each spectrum window the highest signal-to-noise ratio can be obtained without ambiguities caused by either folding or alising (see volume 2 of this series). A pure absorption spectrum improves spectral quality and at the same time enables to discriminate between positive and negative intensities. To obtain phased mode spectra in both dimensions the final signal after 2D FT must contain the terms

$$\mathbf{S\,(f1, f2) = A\,(f1) \cdot A\,(f2) + i \cdot D\,(f1) \cdot D\,(f2)}$$

with no additional "mixed" terms, including both absorptive and dispersive components, e.g. A(f1) · D(f2), left. These mixed terms may be cancelled either by the mode of data acquisition – using phase cycling and/or magnetic field gradients – and/or the Fourier transform procedure itself.

The type of Fourier transformation depends on the type of experiment, the mode of data acquisition and on the way the 2D spectrum should be presented and displayed. In the early days of 2D NMR the magnitude mode was the only way to calculate 2D spectra but today phased mode 2D spectra with pure absorption line shapes in both dimensions is the norm. Magnitude mode 2D spectra are easier to process since no phase corrections need to be done, but have the disadvantage that the cross peaks, calculated from absorptive and dispersive components, have unfavourable line shapes with wide wings in both dimensions. Magnitude mode spectra are no longer preferred even though additional processing (see section 5.3.2) may artificially "narrow" the cross peaks, since this "narrowing" severely affects the signal-to-noise ratio of these spectra. For the details of the various types of 2D FT you are referred to the literature.

With 2D WIN-NMR, 2D Fourier transformation may be accomplished with the commands **xfb**, **xtrf**, **xf2** and **xf1** accessible via the **Process** pull-down menu (Fig. 5.5).

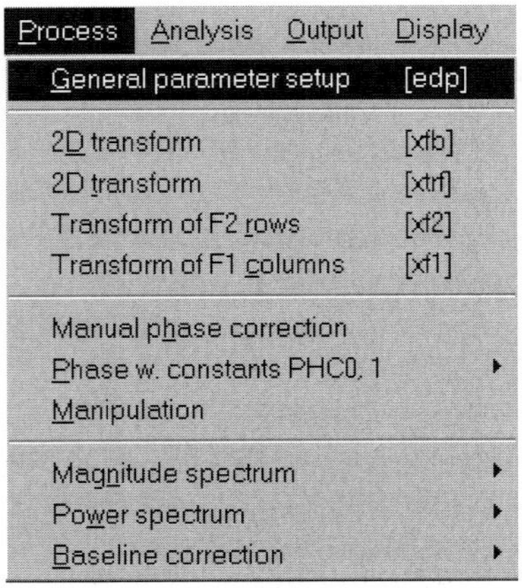

Fig. 5.5: **Process** pull-down menu of 2D WIN-NMR used to initialize a 2D FT.

xfb – 2D Transform in both Dimensions

This command is used to transform the current raw data (SER-file) into a spectrum consisting of real and imaginary parts. The command **xfb** automatically performs the transformation in both the F2 and F1 direction i.e. it executes the two commands **xf2** and **xf1** in sequence but requires less computation time than the separate execution of the two commands. **xfb** executes a sequence of additional commands, depending on the selected processing options and corresponding parameters, accessible in the dialog box (Fig. 5.6) opened by the **General parameter setup** command in the **Process** pull-down menu.The most important of these processing options, baseline corrections in both the time and the frequency domain, linear prediction LP, filtering and phase correction will be discussed in subsequent sections of this chapter.

Hint: If problems arise the **xfb** command, perform a stepwise Fourier transformation using the commands **xf2** and **xf1**.

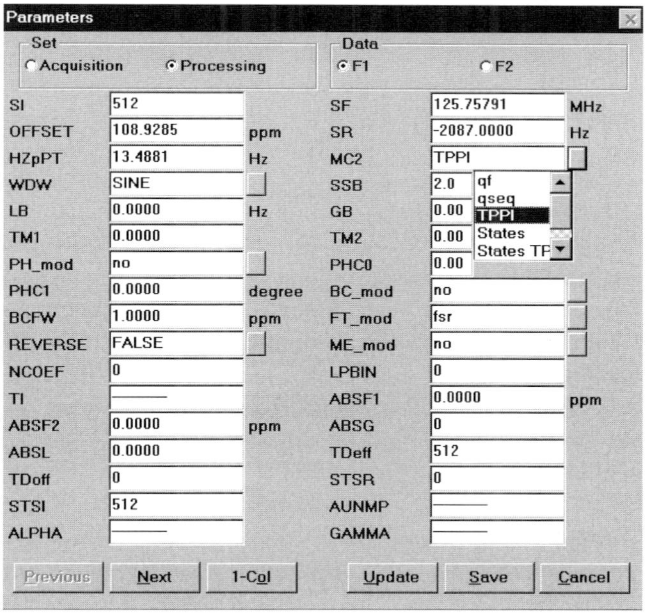

Fig. 5.6: **General parameter setup** dialog box of 2D WIN-NMR. The page for the F1 processing options is shown with a further dialog box opened to select the type of Fourier transformation (MC2).

The type of the Fourier transformation depends upon a series of parameters **AQ_mod**, **FT_mod** and **MC2**, which are usually adjusted either before data acquisition or are automatically adjusted to suit the acquisition parameters. Nevertheless these parameters should be inspected in the **General parameter setup** dialog box and adjusted if necessary, before a 2D FT is initialized. For all the data available in the NMR data base, **AQ_mod** in t2 is **qsim** (the two quadrature detected signals are sampled simultaneously) while **AQ_mod** in t1 is either **qseq** or **qsim**, depending on the type of experiment. **FT_mod** is only used by the **xtrf** command (see below). Through the appropriate choice of the **MC2** parameters the type of transformation is matched to the experimental conditions, **MC2** is automatically set in t2 according to the acquisition parameters but must be set manually in t1 (see below). For all the data available in the NMR data base, **MC2** in t1 is either **qf**, **TPPI** (Time Proportional Phase Increment) or **Echo-Antiecho**. With **qf** a 2D magnitude mode spectrum is obtained and the transformed result is stored in the file with the extensions RR and II of the current data set. With **TPPI** and with **Echo-Antiecho** two different types of 2D FT are applied. A phase sensitive 2D spectrum is obtained and the result is stored in the files with the extensions RR, IR, RI and II.

RR	=	purely real data
RI	=	real in f2, imaginary in f1
IR	=	imaginary in f2, real in f1
II	=	purely imaginary data

For adjusting the **MC2** parameter when processing 2D raw data you are referred to section 5.6 (Tables).

xf2 – 2D Transformation of F2 Rows

This operation performs the 2D-data transformation in the t2 direction only and allows you to inspect your 2D data and to optimize any additional processing before the second FT along t1 is initialized. Again the type of the Fourier transformation depends on the value of the **AQ_mod** and the **MC2** parameters, both of which are automatically adjusted to the acquisition parameters. With **xf2** a series of additional processing steps are sequentially performed, if the corresponding options and parameters have been set. The **xf2** transformation acts on the raw acquisition data (the SER file). The generated data must be processed with **xf1** to complete the 2D transform.

xf1 – 2D Transformation of F1 Columns

This operation performs a transformation in the F1 direction in a similar way to that described for **xf2**. The type of the Fourier transformation depends on the value of the **MC2** parameter, which must be correctly set as described above for the **xfb** command.

xtrf – 2D Transformation in both Dimensions

The command **xtrf** automatically performs a user defined Fourier transformation in both the F2 and F1 dimension. Unlike the **xfb**, the **xf2** and **xf1** commands, **xtrf** takes the processing parameter **FT_mod** into consideration. This option is used for special cases and may be adjusted to your personal needs, e.g. for a real instead of complex FT.

Use the **Help** tool of 2D WIN-NMR for details concerning the **xtrf** command and 2D FT in general.

5.2.5 Phasing of 2D Spectra

Whereas magnitude mode 2D spectra need no phase correction, all phased mode spectra have to be phased in both the F1 and F2 dimension. The phase of the peaks in such a phase-sensitive 2D spectrum depends upon the type of experiment. In phase-sensitive 2D $^1H/^1H$ TOCSY spectra, the inverse detected $^{13}C/^1H$-shift correlated spectra and in the $^1H/^1H$ NOESY spectra, measured for large molecules, all the peaks are in positive absorption. Phase-sensitive $^1H/^1H$ ROESY and $^1H/^1H$ NOESY spectra measured for small molecules display positive cross and negative diagonal peaks while phase-sensitive double quantum filtered $^1H/^1H$ COSY spectra display both positive and negative absorption lines within a single cross or diagonal peak, caused by the passive and active J-couplings respectively. Although several procedures for phasing 2D spectra exist, only one procedure is outlined below.

For 2D data which need no phase correction the **PH_mod** parameter in F1, available in the **General parameter setup** dialog box opened via the **Process** pull-down menu, must be set to **mc** if a magnitude, or to **ps** if a power spectrum should be calculated. Use the **Help** tool for more informations.

Check it in 2D WIN-NMR:

Start 2D WIN-NMR and load the raw data of the 2D magnitude mode COSY spectrum of peracetylated glucose, D:\NMRDATA\GLUCOSE\2D\HH\GHH\001001.SER. From the **Processing** pull-down menu choose the **General parameter setup** item and in the correspondingly opened dialog box inspect the options for **AQ_mod** in the *Acquisition* parameters (F2: **qsim**, F1: **----**). Inspect and eventually modify the *Processing* parameters **MC2** (F2: **----**, F1: **qf**) and **PH_mod** (F2: **no**, F1: **mc**). If you alter one or more parameters click on either the **Update** or **Save** button to quit this dialog box and to take the changes into consideration with the next transformation command, otherwise exit by clicking the **Cancel** button. From the **Process** pull-down menu choose the **xfb** option, start the 2D Fourier transformationand store the spectrum (...\001001.RR).

Repeat these steps but perform a stepwise 2D FT using the **xf2** command first and then the **xf1** command. Inspect the "spectrum" appearing on screen in between the two calculations.

For 2D spectra to be displayed in phased mode, choose the **Manual phase correction** option from the **Process** pull-down menu of 2D WIN-NMR. This will also change the info field of the button panel and disable some of the panel buttons (Fig. 5.7).

To phase either the columns or rows, click on the appropriate button in the button panel. A cross-hair cursor appears in the display field, and the **Ch1** button is highlighted indicating that the trace for the first channel is to be selected. The spectrum display area is internally divided into three parts, corresponding to three different channels (columns or rows) that can be chosen to perform the phase correction. Select appropriate columns or rows from the 2D matrix by moving the cursor to the desired position and clicking with the left mouse button to confirm this position. Activate the next channel by clicking on the **Ch2** or **Ch3** button before selecting the next column or row. Selected traces are extracted from the displayed 2D data and shown in the respective part of the display field.

The chosen trace can differ in appearance from the initially displayed one since the selections proceed on the basis of a reduced data set while the confirmed traces are derived from the full data set. If the confirmed trace is not suitable use the **Prev.Trace** and **Next Trace** buttons to select an appropriate trace.

When all the desired traces have been selected, click the right mouse button to close the selection process. The local mode buttons change and a small upside down triangle marks the position of the automatically computed reference point (biggest point in all of the selected slices). This reference point can be changed by means of the **Big.Point** and **Cursor** buttons. Use the **Help** tool for more information concerning these buttons.

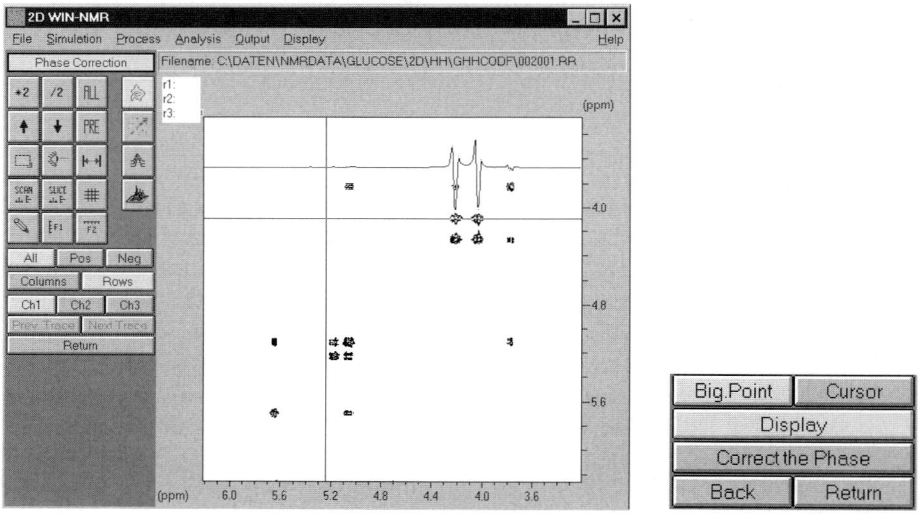

Fig. 5.7: Buttons accessible with the **Manual phase correction** option in the **Process** pull-down menu of 2D WIN- NMR (left). Additional button panel for adjusting the reference point and for switching the 2D display off (right).

One possible method for phasing a 2D spectrum is as follows:

1. Following the instructions given above, select one single row of your spectrum which includes signals as close as possible to both ends of your spectral window. If this is not possible select up to three different rows for this purpose.
2. Terminate the selection of rows by clicking the right mouse button. A new button panel (Fig. 5.7, right) for readjusting the reference point appears. Click on the **Display** button in this new button panel to switch the 2D data display off, permitting a clearer view of the slices during phase correction. Clicking the **Correct the Phase** button results in a further buttons block appearing in the panel (Fig. 5.8). The phase correction can be accomplished in a similar way as with 1D spectra by using the

corresponding scroll bars for adjusting the zero- and first-order phase corrections. Usually, the zero-order phase correction is performed first to obtain the best results for the reference point. For the functions of the other buttons available, **Range+**, **Range–**, **Add 180°**, **Sub 180°** and **Back**, use the **Help** tool.

3. When the phase correction has been completed, click the **Accept** button and a message box appears on the screen. If **Yes** is selected then the **xf2p** command is executed and the F2 phase corrected spectrum will be calculated and shown on the screen. If **No** is selected the new values of the **PHC0** and **PHC1** parameters are stored for future use but no phase correction is performed.
4. If you are satisfied with the result of this correction in F2, select suitable columns for phasing the 2D spectrum in F1. Follow the same procedure as outlined for rows. The **xf1p** command will appear after clicking the **Accept** button to initialize the phase correction in F1.
5. If you are still satisfied with the result, store the spectrum. Since phasing of 2D spectra may be time-consuming store the evaluated correction values **PHC0** and **PHC1** for F2 and F1 together with the raw data, using the **Save** button in the **General parameters setup** dialog box. The next time you want to calculate your 2D spectrum with the **xfb** or the **xf2** and **xf1** commands using these phase correction values, simply set the **PH_mod** parameter in the **General parameter setup** dialog box to **pk** for both dimensions. This will automatically perform a phase correction in F2 and F1 using the stored values **PHC0** and **PHC1**.

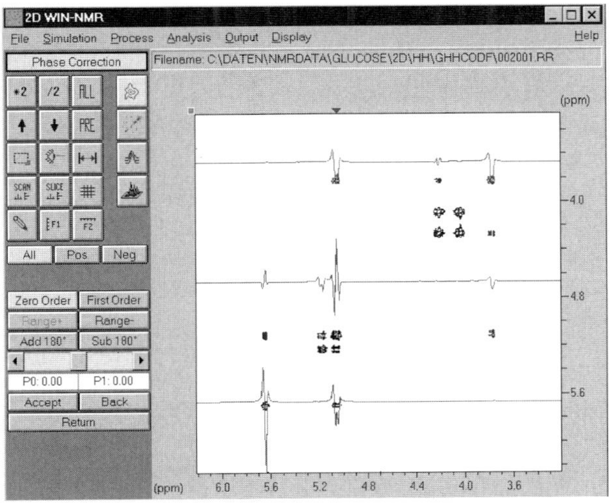

Fig. 5.8: Buttons for phasing 2D spectra available with the **Correct the Phase** button.

The type of phase correction required depends on the nature of the 2D experiment. Fig. 5.9 shows some typical phase corrected rows obtained from a number of popular phase sensitive 2D experiments. The same phasing behaviour is valid for the corresponding columns:

- **$^1H/^1H$-double quantum filtered COSY**: Note that with this experiment both the diagonal and the cross peaks may be phased to pure absorption. Therefore it is best to select diagonal peaks at the extremes and in the center of the spectrum for phase adjustment. The cross peaks when correctly phased consist of positive and negative peaks, which are anti-phase with respect to the active, and are in-phase with respect to the passive coupling(s).
  Note that with a non-DQ-filtered, phase sensitive COSY experiment the cross peaks are again purely absorptive while diagonal peaks irrespective of the phase correction will have both absorptive and dispersive character. Unlike most other 2D spectra, it is therefore best to phase correct a non-DQ-filtered phase sensitive COSY spectrum while examining the cross rather than the diagonal peaks.
- **$^1H/^1H$-TOCSY**: Both cross peaks and diagonal peaks should appear in positive absorption after proper phasing. Use the diagonal peaks to correct the phase.
- **$^1H/^1H$-ROESY (-NOESY)**: Cross peaks and diagonal peaks appear in absorption. The diagonal peaks are the most intense and are best suited for phase adjustments. They should be phased for negative absorption in ROESY spectra and in NOESY spectra measured for small molecules, giving cross peaks in positive absorption in both cases. For NOESY spectra of large molecules (e.g. biomolecules), both diagonal and cross peaks should be phased to positive absorption.
- **$^{13}C/^1H$-Shift Correlation**: For most heteronuclear phase sensitive 2D spectra, all cross peaks should appear in positive absorption. Exceptions are spectra obtained with DEPT modified experiments, designed to discriminate between different carbon multiplicities via positive and negative cross peak intensities.

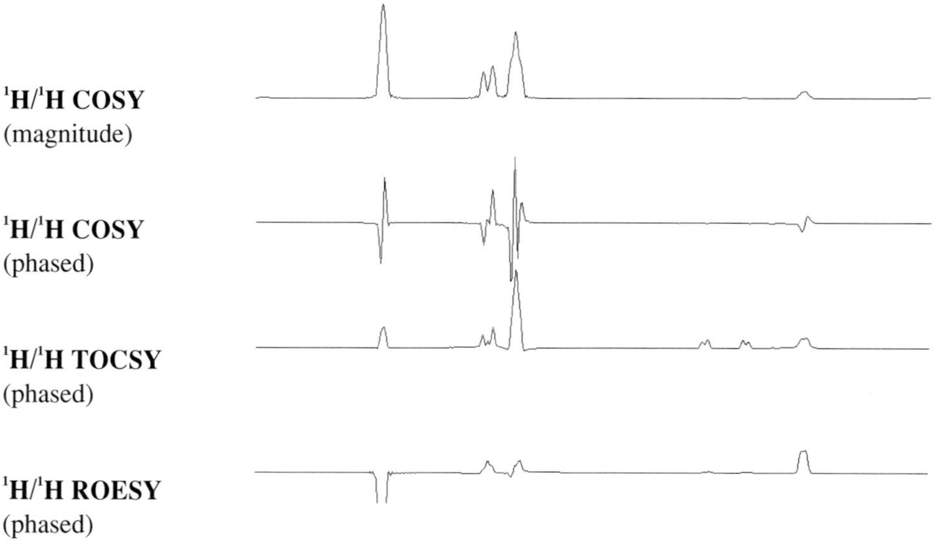

Fig. 5.9: Examples of correctly phased rows from some phase sensitive 2D experiments.

Check it in 2D WIN-NMR:

Load the raw data obtained for peracetylated glucose with the 2D TOCSY experiment D:\NMRDATA\GLUCOSE\2D\HH\GHHTO\001001.SER and perform a 2D FT following the guidelines given above. Enter the **Manual phase correction** option in the **Process** pull-down menu and perform a phase correction in F2 and F1 according to the procedure outlined above. Try to phase all peaks to positive absorption and store the spectrum (...\001001.RR).

Check it in 2D WIN-NMR:

Load the raw data obtained for peracetylated glucose with the 2D ROESY experiment D:\NMRDATA\GLUCOSE\2D\HH\GHHRO\001001.SER and perform a 2D FT following the guidelines given above. Enter the **Manual phase correction** option and perform a phase correction in F2 and F1 according to the procedure outlined above. Use the intense diagonal peaks for the phase adjustments, phase them to negative absorption and inspect the phase of the cross peaks, which should now appear in positive absorption. Try to explain why a few of them still have both absorptive and dispersive components.Store the phase-corrected spectrum (...\001001.RR).

Check it in 2D WIN-NMR:

Load the raw data obtained for peracetylated glucose with the 2D double quantum filtered COSY experiment D:\NMRDATA\GLUCOSE\2D\HH\GHHCODF\001001.SER and perform a 2D FT following the guidelines given above. Enter the **Manual phase correction** option and perform a phase correction in F2 and F1 according to the procedure outlined above. Use suitable diagonal peaks for the phase adjustment. Try to get cross peaks with both negative and positive signals and store the spectrum(...\001001.RR). From the **Process** pull-down menu choose the **General parameters setup** option. In the dialog box set the **pk** options for F2 and F1 and check the correction factors (**PHC0**, **PHC1**) for F2 and F1 which should still be the same as adjusted above. Click the **Save** button to save the setting and to exit the dialog box. Execute **xfb** with the same raw data, which will perform the Fourier transformation and the phase correction in both dimensions.

Check it in 2D WIN-NMR:

Load the raw data obtained for peracetylated glucose with the inverse 2D CH-COSY experiment D:\NMRDATA\GLUCOSE\2D\CHI\GCHICOMQ\001001.SER and perform a 2D Fourier transformation following the guidelines given above. Enter the **Manual phase correction** option and perform a phase corrections in F2 and F1 according to the procedure outlined above. Try to phase all peaks for positive absorption and store the spectrum (...\001001.RR).

5.3 Advanced Processing in the Time Domain

5.3.1 Introduction

There exist a set of additional processing tools designed to improve the quality of measured raw data (FID) and to enhance the final spectrum for subsequent spectral analysis. In general the processing of 1D or 2D data sets obeys the same general rules, although both types of data have options specific to them. Any time domain signal, e.g. the NMR raw data s(t), may be Fourier transformed and together with its frequency domain counterpart S(f) forms a Fourier pair (Fig. 5.10). The two functions are mutually connected by the Fourier transformation and the inverse Fourier transformation respectively.

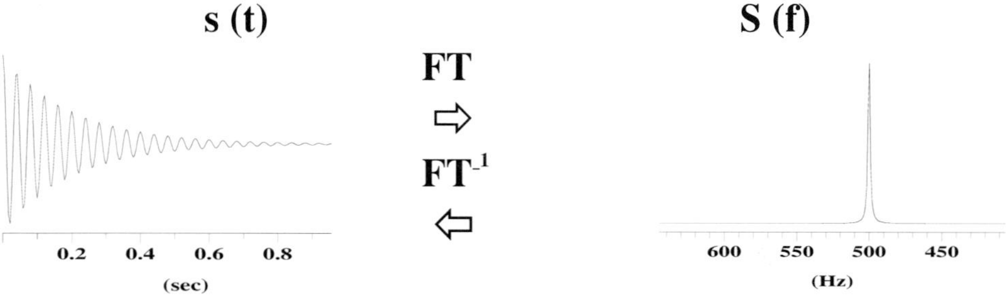

Fig. 5.10: The FID and the spectrum form a Fourier pair.

Processing in general transforms an original time domain signal s(t) with the aid of some processing function(s) into a manipulated frequency domain signal S''(f). Manipulations can be performed either in the time domain (**s(t)** ⇨ **s'(t)**) prior to the Fourier transformation, or in the frequency domain (**S'(f)** ⇨ **S''(f)**) after the Fourier transformation as illustrated in Fig. 5.11.

FT

s (t) ⇨ **s' (t)** ⇨ **S' (f)** ⇨ **S'' (f)**

Fig. 5.11: The basic steps to process NMR data.

NMR data is usually processed using one or more processing functions, some of which are applied in the time domain, others in the frequency domain. Each processing function in the time domain f(t) also has its counterpart F(f) in the frequency domain and forms a Fourier pair. In principle the same effect in the final spectrum S''(f) may be obtained with a given processing function, applied either in the time or the frequency domain as long as a few important rules are followed when performing the corresponding algebraic operations. In practice and for simplicity reasons, however, most of the processing is usually performed on the raw data (FID), prior to the Fourier transformation.

With s(t) and f(t) beeing the FID of an NMR signal and a processing function or a second FID respectively and with S(f) and F(f) beeing the corresponding frequency domain counterparts, the following rules govern the manipulation of these Fourier pairs:

Time Domain:	Frequency Domain:
s(t) +/– f(t)	**S(f) +/– F(f)**
a · s(t)	**a · S(f)**
s(t) · f(t)	**S(f) × F(f)**

Adding/subtracting two time domain functions s(t) and f(t) is equivalent to adding the two corresponding frequency domain counterparts. This equivalence allows to obtain e.g. an NOE difference spectrum either by subtracting the two corresponding FIDs or by subtracting the two corresponding spectra.

The multiplication of the FID s(t) with a factor a is equivalent to a multiplication of the spectrum S(f) with the same factor. This equivalence may be exploited e.g. in spectral editing to obtain multiplicity selective ^{13}C subspectra from the DEPT-45, DEPT-90 and the DEPT-135 data. The multiplication with the corresponding factors may be performed with the DEPT FIDs or the DEPT spectra.

The multiplication of an FID s(t) with a second time domain function f(t) is not simply the multiplication, but the convolution of the two corresponding frequency domain counterparts. Since a multiplication is much simpler to perform – even for a computer – correction functions to improve spectral quality are almost exclusively applied in the time domain (see below).

Some representative time domain signals (FID) obtained with 1D and 2D experiments are shown in Figs. 5.12 and 5.13.

The examples in Figures 5.12 and 5.13 may be described as follows:

1. FIDs in general are composed of NMR signals and noise. The decay of the NMR signals depends on the transversal spin-spin relaxation time T_2 (or - to be precise - T_2*, the effective transversal relaxation time taking into account inhomogeneities of the static magnetic field) of the nuclei involved. Signals of nuclei with short T_2 decay faster compared to signals of nuclei with long T_2. The noise level is more or less constant throughout the whole acquisition time. Spectrometer imperfections or instabilities may give rise to additional "noise" and in most cases (pulse breakthrough, acoustic ringing) manifests itself in the very first part of the FID. As a consequence, the signal-to-noise ratio of an FID decays throughout the acquisition time (Fig. 5.14). This fact is exploited in NMR data processing to enhance the signal-

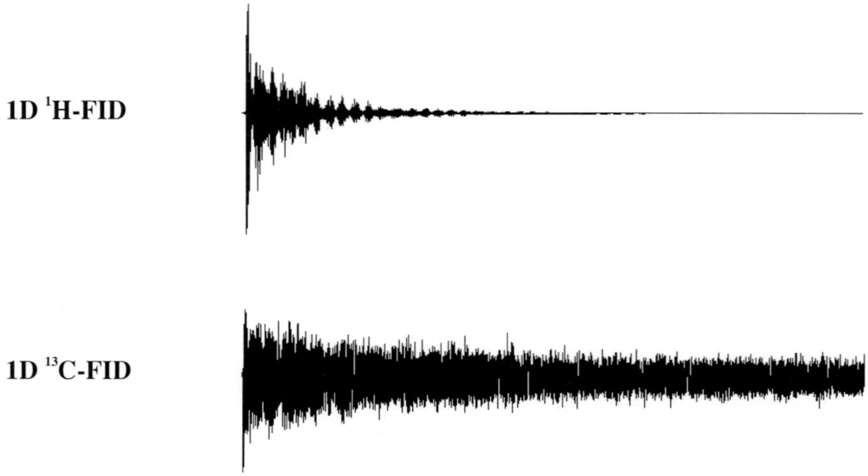

Fig. 5.12: Representative FIDs obtained with 1D experiments.

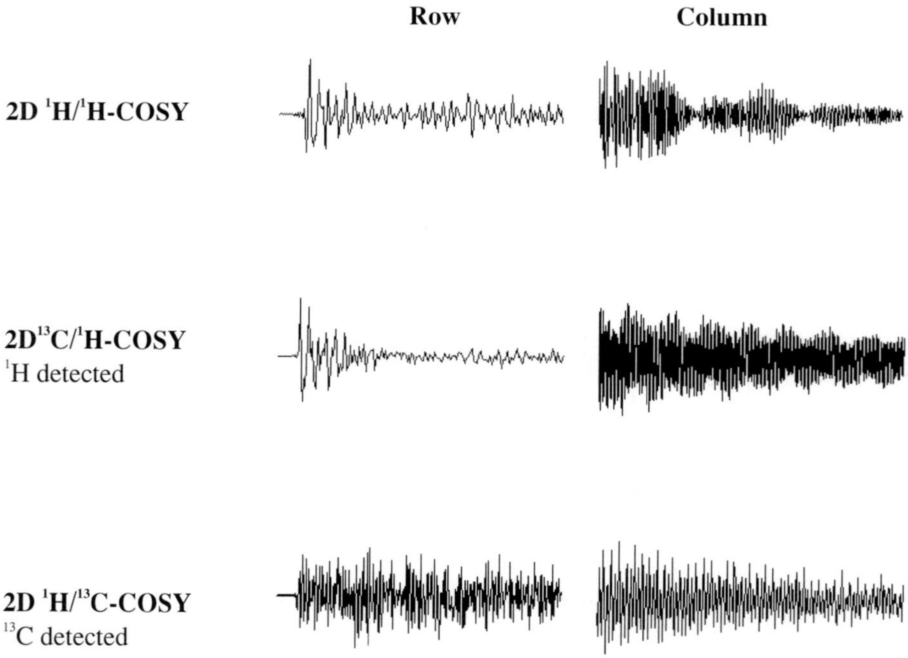

Fig. 5.13: Representative FIDs obtained with 2D experiments.

to-noise ratio of an FID decays throughout the acquisition time (Fig. 5.14). This fact is exploited in NMR data processing to enhance the signal-to-noise ratio, to improve the spectral resolution and to suppress effects caused by unwanted "spectrometer noise". Emphasizing the first part of the FID improves the signal-to-noise ratio, but makes the resolution worse, since the sharp signal components with long T_2 are suppressed with respect to broad signal components with short T_2. On the other hand emphasizing the second part of the FID improves the spectral resolution, suppresses or even removes the unwanted effects of many artifacts, but at the same time decreases the signal-to-noise ratio. Different procedures must therefore be applied depending on the type of NMR data and on your requirements.

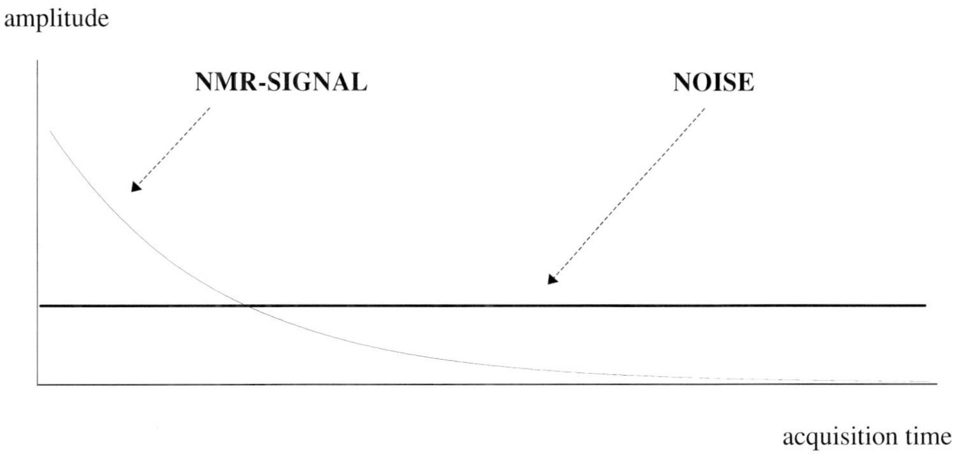

Fig. 5.14: NMR signal and noise levels in an FID.

2. Although at first glance all FIDs look similar, the shape of FIDs differs and depends on the sample amount, kind of experiment (1D, 2D), the nuclei detected (^1H, ^{13}C) and on acquisition parameters such as the length of the acquisition time, the spectral window and the number of time domain data points TD. Subsequent processing must take this into account and be tailored to these different shapes.
3. FIDs measured for highly sensitive nuclei (^1H) usually show a very high signal-to-noise ratio, in contrast to low-sensitive nuclei (^{13}C), for which "noisy" FIDs are obtained.
4. For FIDs obtained with 1D experiments, the signal has usually decayed close to zero, since the acquisition time is large compared to the relaxation time T_2. However the FIDs (both rows and columns) obtained in a 2D experiment have usually not completely decayed within the available acquisition times in t2 and t1 respectively, since the acquisition times are short compared to the relaxation time T_2.
5. The signal-to-noise ratio of rows and columns obtained with 2D experiments vary. They are often different for rows and columns and depend on the mode of detection (^1H/^{13}C-detected), on the kind of experiment and its experimental parameters and on the sample amount.

The main consequences, problems and the processing required for the different types of FIDs are as follows:

1. FIDs should generally be processed prior to Fourier transformation to improve spectral quality.
   ⇨ **Spectral improvement for subsequent spectral analysis**
2. For 1D ^{1}H-spectra the signal-to-noise ratio is usually very high, and the main problems concern the resolution of individual signals in overcrowded spectral regions.
   ⇨ **Resolution enhancement**
3. 1D spectra of low-sensitive nuclei are usually well resolved due to the larger range of chemical shifts. ^{1}H broadband decoupling during acquisition simplifies the spectrum further yielding singlet rather than multiplet signals. However, the signal-to-noise ratios, especially with small sample amounts, are often low.
   ⇨ **Sensitivity enhancement**
4. With 2D spectra, depending on the kind of experiment, problems with resolution and/or sensitivity may arise. Resolution is an inherent problem with 2D experiments and is mainly caused by the limited amount of spectrometer time and disk storage capacity. Sensitivity may be a problem when measuring small sample amounts or weak effects (NOESY, ROESY, ...). In heteronuclear 2D experiments the sensitivity problem is more pronounced with X- rather than with ^{1}H detection.
   ⇨ **Resolution and/or Sensitivity enhancement**
5. When non-decayed FIDs are directly Fourier transformed, truncation-effects resulting in line distortions occurs. This problem is most pronounced with 2D data.
   ⇨ **Elimination of truncation effects**
6. Incorrectly set experimental parameters or spectrometer problems/instabilities may cause unwanted effects (baseline-roll, spikes, ...).
   ⇨ **Filtering or suppression of spectral artifacts in the spectra**

A whole series of processing steps is usually performed to overcome these problems and to produce high quality spectra. The way raw data is processed is highly dependent on the spectroscopic problem itself and is therefore tailored to the NMR parameter(s) required for structural elucidation:

- Sensitivity enhancement is applied to recognise weak signals of slow relaxing, quaternary carbons in noisy ^{13}C spectra, or to measure the weak ^{1}H signals of low concentration impurities.
- Resolution enhancement on the other hand is applied to resolve overcrowded spectral regions or to measure very small signal splitting caused by J-coupling.
- Adding blocks of separately acquired FIDs together is recommended when studying unstable compounds or when using difference techniques (NOE-difference, decoupling difference). In the former case, coaddition of blocks of data allows snap-shots to be taken of the sample decomposing whilst in the latter case it minimizes effects caused by spectrometer instability and so improves the visibility of the very small difference effects.

However the choice of processing options is not only governed by these problems, the spectral quality and the parameter of interest, but must also take into account the mode of Fourier transformation (magnitude-/phased-mode). Last but not least, processing depends on your preferences, since the degree to which the final spectrum is affected by the various processing steps is very different and in most cases alternative processing exist for reaching the same goal.

Processing in general may be classified according to its purpose and its effect on the final spectrum (e.g. sensitivity enhancement) or may be classified according to the principles involved. Whereas the purposes for processing have been outlined above, the main principles may be described as follows:

1. There exist two main types of time-domain processing options: the FID s(t) may either be added to or may be multiplied (weighted) by a processing function f(t). Multiplication of an FID with a processing function f(t) may affect all or only part of the FID, whereas the addition of a processing function f(t) to an FID may retain or increase the number of acquired data points $TD_{aq.}$ (Fig. 5.15). Shifting the original FID along the time axis represents a further principle for data manipulation of minor importance.

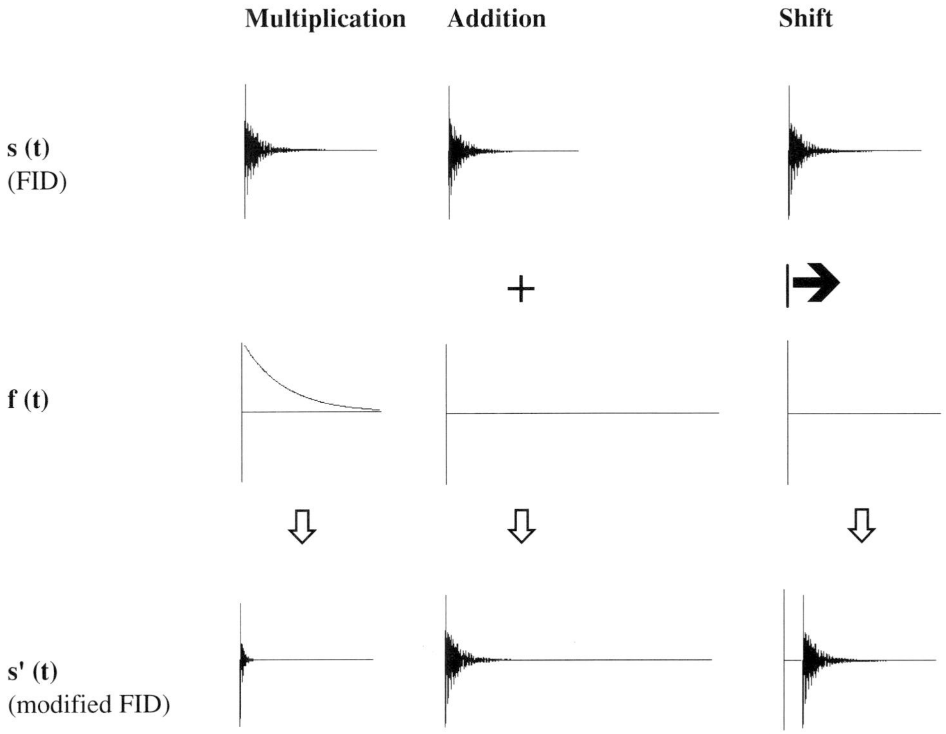

Fig. 5.15: Different principles for time domain processing of FIDs: s(t) and s' (t) is the acquired and the modified FID respectively, f(t) is the processing function.

2. Processing options based on different principles but serving the same purpose exist.
3. Processing functions are either dedicated to one single purpose (e.g. the suppression of "spectrometer artifacts") or serve different purposes (signal-to-noise or resolution enhancement) depending on the corresponding function parameter(s) (see below).
4. The original FID s(t) may be treated with several time-domain processing functions one after the other. The order in which the individual processing steps are performed is arbitrary and does not influence the final spectrum.

Both 1D WIN-NMR and 2D WIN-NMR have various tools available for processing NMR raw data in the time domain. In the following discussion, these tools are presented in the same order as usually applied in daily work. The effects of various processing options will be demonstrated and their field of application, their advantages and limits are briefly discussed. *Check its* allow you to become familiar with these tools and to experience their great value in optimizing spectral quality and adjusting your spectra to your needs. A table containing recommendations for processing options and their corresponding parameters for the most popular 1D and 2D experiments is given in section 5.6.

Use the **Help** tools available with 1D- and 2D WIN-NMR for a detailed description of the various processing options and of their application.
Processing of time domain data is accessible after having loaded either a 1D (<NAME>.FID) or 2D (<NAME>.SER) raw data file using the corresponding **Processing** pull-down menus (Fig. 5.16).

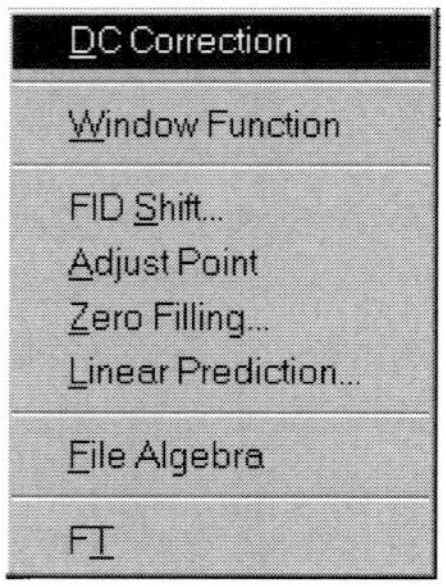

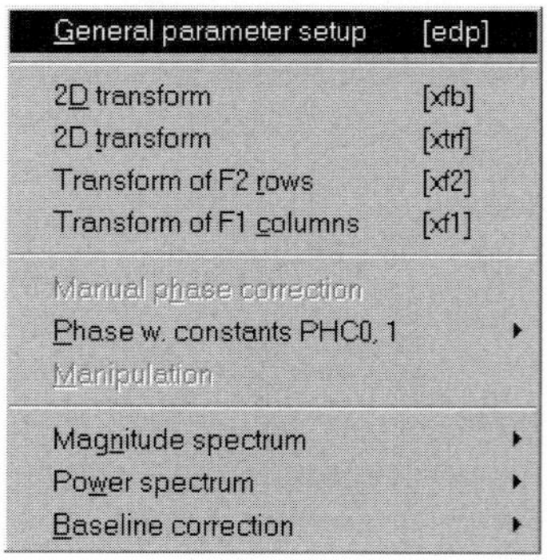

Fig. 5.16: **Process** pull-down menus available with 1D WIN-NMR (left) and 2D WIN-NMR (right).

For 1D WIN-NMR, a number of processing options can be selected from the **Process** pull-down menu and the parameters related to some of these options may be adjusted interactively. 2D WIN-NMR offers a general parameter display with several edit fields, which allow you to directly set the processing option and its associated parameters. Furthermore, individual rows or columns of 2D data may be transferred into 1D WIN-NMR for additional processing utilizing the more powerful processing options of this program.

5.3.2 Multiplication with a Processing Function: s(t) · f(t) "Weighting", "Filtering", "Apodization"

1D WIN-NMR offers a wide range of window functions, with their corresponding parameters, for multiplying the FID. These functions and parameters may be inspected or directly executed using the **Interactive** and the **Execute** button respectively in the **Specify Window Function** dialog box (Fig. 5.17) accessible with the **Window Function** option in the **Process** pull-down menu.. Furthermore, a **Window!** button in the button panel is available whenever an FID is loaded and is the simplest and quickest way of processing an FID. If the window function and its parameter has been optimized on a previous data set, weighting can directly be initialized by clicking with left mouse button on the **Window!** button. Clicking with the right mouse button on the **Window!** button opens the same dialog box (Fig. 5.17) and allows to readjust the parameters prior to start the weighting process.

The corresponding menu for 2D data to set up the window functions and their parameters in F2 (rows) and F1 (columns) is shown in Figs. 5.18 and 5.19. Processing parameters for both F2 and F1 may both be set prior to the consecutive Fourier transformation in both dimensions started with the **xfb** command. Alternatively the processing and Fourier transformation in t2 – initialized with the **xf2** command – may be performed first. Suitable columns may then be used to adjust and set the processing parameters in t1 prior to the second Fourier transformation, started with the **xf1** command.

The available window functions (with their corresponding parameters) show different shapes which may be modified and tailored to the actual data and to the subsequent spectra analysis.

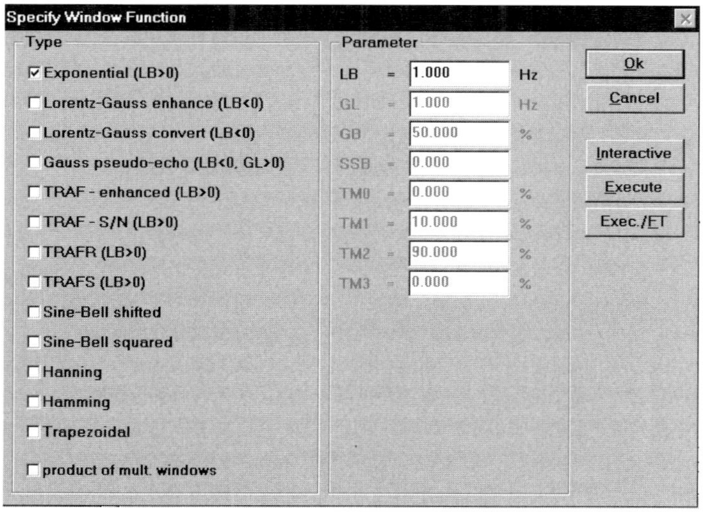

Fig. 5.17: Window functions available with 1D WIN-NMR.

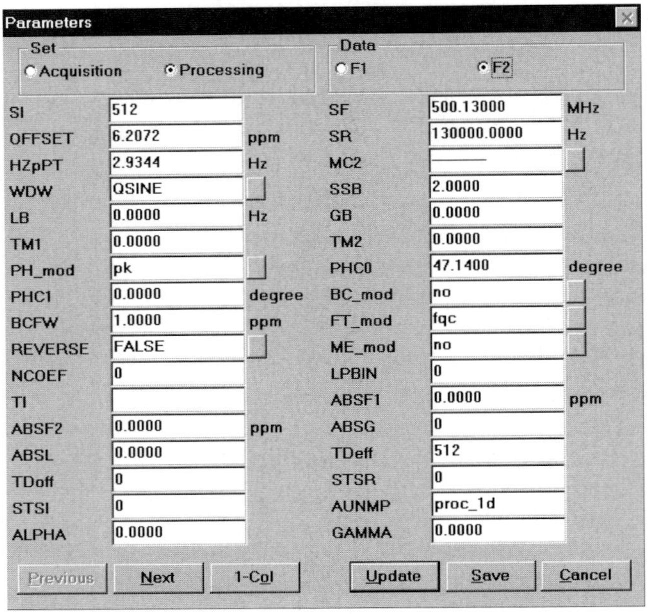

Fig. 5.18: Processing options and corresponding parameters available with 2D WIN-NMR. The menu for processing in F2 is shown.

The available window functions (with their corresponding parameters) show different shapes which may be modified and tailored to the actual data and to the subsequent spectra analysis.

The **Hamming** and the **Hanning** windows available with 1D WIN-NMR have no parameters and are used exclusively for signal-to-noise enhancement. All other windows use either one (**Exponential**, **Lorentz/Gauss convert**, **TRAF-enhanced**, **TRAF-S/N**, **TRAFR**, **TRAFS**, **Sine-Bell shifted, Sine-Bell squared**), two (**Lorentz-Gauss enhance**), three (**Gauss pseudo-echo**) or even four (**Trapezoidal**) parameters for adjusting the window shape. All these parameter defined windows are bifunctional and may be used either to suppress the first or the second part of the FID thereby enhancing the signal-to-noise ratio or improving the resolution respectively.

In principle several types of window functions can also be combined with the **product of mult. windows** option in the **Specify Window Function** dialog box.

In the following *Check its*, dedicated to illustrate the effect of various window functions and to study the influence of the corresponding parameters, store the processed spectra under the same name, but with ascending processing numbers (<NAME>\001001.1R, <NAME>\001002.1R, ...). Use the dual/multiple display of 1D WIN-NMR to display and plot the whole series. Do not apply the **DC Correction** at this stage. Use the **Help** tool for the mathematical description of the various window functions and for more details of how to use them.

Data files of different spectral quality have been acquired under special experimental conditions (Table 5.1), to best demonstrate the effects obtained with the various weighting functions.

Table 5.1: Data files to best demonstrate the effects obtained with weighting

Data file: /GH	Spectrum range	TD	Spectral quality	Aim
004001	full	64K	HR[1)]	reference
005001	full	64K	MR[2)]	resolution enhanced
006001	full	64K	LR[3)]	resolution enhanced
007001	expansion	16K	HR	reference
008001	expansion	16K	MR	resolution enhanced
009001	expansion	16K	LR	resolution enhanced
010001	full	64K	HR	reference
011001	full	64K	MR	resolution and S/N enhanced
012001	full	64K	LR	resolution and S/N enhanced

[1)] : High Resolution (HR); sample spinning, homogeneity optimized
[2)] : Medium Resolution (MR); sample not spinning, homogeneity optimized
[3)] : Low Resolution (LR); sample not spinning, homogeneity not optimized

Check it in 1D WIN-NMR:

Load the ^{13}C FID of peracetylated glucose D:\NMRDATA\GLUCOSE\1D\C\GC\001001.FID and Fourier transform the data. Phase the spectrum and store it under the same name (reference spectrum). Recall the FID and apply an exponential window. From the **Process** pull-down menu choose the **Window Function** option and in the dialog box select the **Exponential** window type. Use the **Interactive** option to adjust and set the corresponding parameter **LB** (line broadening). Note that exponential weighting with **LB** > 0 emphasizes the signal-to-noise ratio with respect to resolution. Try out different **LB** values and adjust it for the best compromise (signal-to-noise versus resolution). Store the corresponding spectra under ascending processing numbers (001002, 001003, ...) and choose a spectral region with closely spaced signals, e.g. the carbonyl region, for this optimization.

Try out the **Window!** button in the button panel for rapid processing. Note that when clicking with the right mouse button on the **Window!** button the window dialog box appears. This allows to most conveniently inspect/adjust the weighting parameters before this processing step is initialized with the left mouse button.

Determine the mean linewidth at half height $\Delta\upsilon_{1/2}$, in the spectrum obtained with no weighting applied, calculate the parameter $LB_{matched}$ for the matched filter according to : $LB_{matched} = \pi \cdot \Delta\upsilon_{1/2}$ and then apply it. The matched filter should yield the best compromise with respect to signal-to-noise and resolution. Inspect and compare the signal-to-noise ratio and the resolution for the various **LB** values and store the best compromise as ...\001001.1R.

Check it in 1D WIN-NMR:

Load the ^{13}C FID of peracetylated glucose D:\NMRDATA\GLUCOSE\1D\C\GC\001001.FID and apply the four **Traficante** window types. In this case, use the **Interactive** option to enhance the signal-to-noise ratio. Use different values for **LB** and compare the results. Compare the results with the unweighted and the exponentially weighted spectra.

Check it in 1D WIN-NMR:

Load the ^{1}H FID of peracetylated glucose D:\NMRDATA\GLUCOSE\1D\H\GH\004001.FID and Fourier transform the data without applying any window function. Phase the spectrum and store it (reference spectrum). Recall the FID and apply an exponential window. From the **Process** pull-down menu choose the **Window Function** option and in the dialog box select the **Exponential** window type. Use the **Interactive** option to adjust and set the corresponding parameter **LB** (line broadening). Note that exponential weighting with **LB** > 0 emphasizes the signal-to-noise ratio with respect to resolution. Use different **LB** values in the range of 0.1 - 4 Hz and compare the results with the reference spectrum using either the dual or the multiple display. What can be concluded with respect to the resolution and the signal-to-noise ratio and what can be concluded with respect to the corresponding results with the ^{13}C data?

Check it in 1D WIN-NMR:

Load either the ^{1}H FIDs D:\NMRDATA\GLUCOSE\1D\H\GH\004001-006001.FID (full spectrum) or D:\NMRDATA\GLUCOSE\1D\H\GH\007001-009001.FID (expansion) of peracetylated glucose and Fourier transform the data. Phase the spectra in the same way and store them (reference spectra). Inspect and compare the three spectra with respect to resolution. Try to improve the resolution for those spectra with medium or low resolution (see Table 5.1). Apply a **Sine-Bell shifted** and a **Sine-Bell squared** window type following the same procedure as with the exponential window. Again there is a single parameter **SSB**, which is applied to shift (and to stretch) the sine-bell window with respect to the FID and which is used to emphasize either the first or the second part of the FID thereby optimizing the signal-to-noise ratio or the resolution respectively. Use different **SSB** values and compare the results with the reference spectra. Note that for **SSB** < 2 an unshifted sine-bell window results, whereas for **SSB** = 2 the window function shows a maximum for t = 0. With increasing values (**SSB** > 2) this maximum moves to larger t values. As a consequence a maximum signal-to-noise improvement is expected for **SSB** = 2, whereas the best resolution should result with **SSB** = 0. Which additional effect occurs with **SSB** = 0 and why? Use a few values for **SSB** and compare the corresponding results obtained with a **Sine-Bell shifted** and a **Sine-Bell squared** window.

Check it in 1D WIN-NMR:

Use the same series of data and follow the same procedure as before to try out the four **Traficante** window types. Use the **Interactive** option, as a starting value set **LB** = 0, which yields a horizontal line and corresponds to no weighting. Increase and decrease the **LB** value in small steps using the scroll bar and try to predict the effect on your spectrum. Use different values to enhance the resolution, store the results and compare the resulting spectra using the multiple display.

Check it in 1D WIN-NMR:

Load the ^{13}C FID of peracetylated glucose C:\DEMO1D\DAT\ASP3000\CDIS\001001.FID originally acquired in the DISNMR format, but converted into the WIN-NMR format in chapter 2.7.2.2, and process it to obtain the best spectral quality.

Check it in 1D WIN-NMR:

Use the same series of data and follow the same procedure as before to try out the **Lorentz-Gauss** convert window type. There is one single parameter **LB** available to adjust the window. Set the initial value to **LB** = 0, increment and decrement its value in small steps and inspect the shape of the window using the interactive mode. Note that for **LB** > 0 the shape of the window is similar to the exponential window (signal-to-noise improvement) whereas for **LB** < 0 the window shape is similar to the sine-bell squared window. Try out a few values to enhance the signal-to-noise ratio and to improve the resolution, store the results and compare the spectra using the multiple display.

Check it in 1D WIN-NMR:

Use the same series of data and follow the same procedure as before again to try out the **Lorentz-Gauss enhance** window type. There are two parameters (**LB**, **GB**) available in this case. Set **GB** = 100 and **LB** = 0 as the initial values, increment and decrement the value of **LB** in small steps and inspect the shape of the window using the **Interactive** option. Note that for **LB** > 0 the shape of the window is similar to the exponential window (signal-to-noise improvement) whereas for **LB** < 0 the window shape is similar to a shifted sine-bell squared window (resolution enhancement) with its maximum at the end of the FID. With **LB** < 0, decrease **GB** in small steps towards 0.

This will shift the maximum of window close to the beginning of the FID. Try out a few values to enhance the signal-to-noise ratio and to improve the resolution. Note that with two parameters to be adjusted processing becomes more flexible, but also more time consuming.

Check it in 1D WIN-NMR:

Use the **Gauss pseudo-echo** window type with the same data set and try out the three window parameter **LB**, **GB** and **GL**. Use the **Interactive** option and calculate the spectra for a few parameter combinations.

Check it in 1D WIN-NMR:

Load now the ^{1}H FIDs of peracetylated glucose D:\NMRDATA\GLUCOSE\1D\H\GH\010001-012001.FID and Fourier transform the data. Phase the spectra and store them (reference spectra). Note that in this case the signal-to-noise ratio is low for all three spectra and that resolution is not the best for the second and third data set (see Table 5.1). Try to find the best compromise with respect to signal-to-noise ratio and resolution. Use different weighting functions for this optimization and store the results separately. Compare and interpret the results.

Check it in 1D WIN-NMR:

Load the ^{1}H FID of peracetylated glucose C:\DEMO1D\DAT\ASP3000\HDIS\001001.FID originally acquired in the DISNMR format and process it to obtain the best spectral quality.

With 2D WIN-NMR the type of window and its parameters are selected in the **Parameters** dialog and edit box (Fig. 5.18). There are five windows available (Fig. 5.19) which must be defined for both the F2 and F1 dimension.

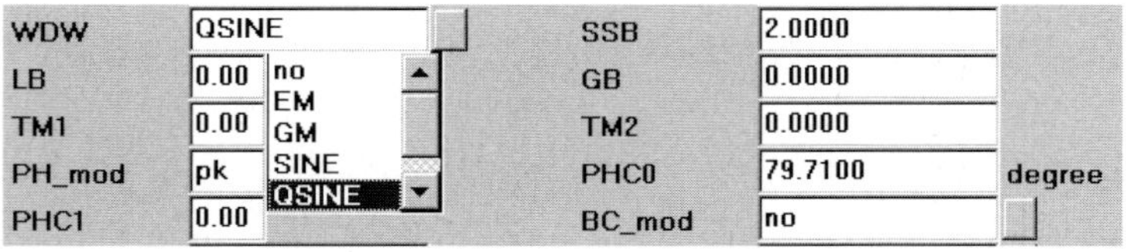

Fig. 5.19: **WDW** window box activated with the **WDW** button from within the **Parameters** dialog box

Check it in 2D WIN-NMR:

Load the raw data of the gradient enhanced ^{1}H/^{1}H COSY (magnitude mode) experiment D:\NMRDATA\GLUCOSE\2D\HH\GHHCO\002001.SER and use either the sine (**SINE**) or the sine squared (**QSINE**) window for weighting in both either the sine (**SINE**) or the sine squared (**QSINE**) window for weighting in both dimensions. Note that in view of the subsequent magnitude calculation, an unshifted window (**SSB** = 0) is usually applied. This decreases the overall sensitivity (which in most cases is not a problem with $^1H/^1H$ COSY data) but also removes the wide wings introduced with the magnitude mode, which would otherwise severely affect the spectral resolution. Check this by setting **SSB** = 2 for a second calculation. For F1 set **PH_mod** to **mc** (magnitude calculation) before the Fourier transformation for an automatic magnitude calculation. Store the differently weighted 2D spectra with different processing numbers. Select a suitable row and a suitable column and transfer the corresponding slices of each 2D spectrum into 1D WIN-NMR using the **Slice** button in 2D WIN-NMR. Store each of these slices and then use the dual display mode to compare the effect (signal-to-noise ratio, resolution) of different window types (e.g. sine and sine squared) and of different **SSB** parameters (0/2).

Note: If the FT in t1 has been applied with **PH_mod** for F1 accidentally set to **no**, choose the **Magnitude spectrum** submenu from the **Process** pull-down menu and select the command **of F1 columns [xf1m]** for a subsequent magnitude calculation.

Check it in 2D WIN-NMR:

Load the raw data of the gradient enhanced, double quantum filtered, phase sensitive $^1H/^1H$ COSY experiment D:\NMRDATA\GLUCOSE\2D\HH\GHHCODF\002001.SER. Again use the sine (**SINE**) or the sine squared (**QSINE**) window in both dimensions. Note that in this case with no magnitude calculation applied, the windows are shifted (**SSB** = 2) to improve the signal-to-noise ratio and to prevent signal distortions. However in F1 **SSB** may be set to **SSB** = 0 or **SSB** > 2 in order to emphasize small couplings. Process the spectrum with **SSB** = 2 in both dimensions, phase the spectrum and store it. Process the spectrum again, but set **SSB** = 0 for F1. Compare suitable rows and columns of these spectra (e.g. row 372, column 372) with respect to the signal-to-noise ratio, the ratio of small to large couplings and the line shapes. Use the **Scan** and **Slice** options in 2D WIN-NMR and the dual display in 1D WIN-NMR to select, transfer and inspect corresponding rows and columns.

Check it in 2D WIN-NMR:

Load the raw data of the gradient enhanced inverse detected $^{13}C/^1H$ COSY (HMQC) experiment D:\NMRDATA\GLUCOSE\2D\CH\GCHICOMQ\ 001001. SER. Apply different window combinations, e.g. **QSINE**, **SINE**, **EM** in both dimensions and try out different window parameters. Note that for heteronuclear 2D spectra, especially weak samples, the signal-to-noise ratio is the crucial factor. As a consequence the signal-to-noise ratio is usually optimized, i.e. **SSB** is set to 2 for **QSINE** and **SINE** windows and **LB** is adjusted accordingly for **EM**. However, in a few cases, and if the signal-to-noise ratio is high enough, the resolution may be improved in one or even both dimensions, with the window parameters set accordingly, to resolve overcrowded regions in the carbon or the proton domain. Phase the spectrum after Fourier transformation. Store the differently weighted spectra and again use the **Scan**, **Slice** options and the dual/multiple display in 1D WIN-NMR to select, transfer and inspect suitable rows and columns of these spectra.

Improving the resolution in F2 (^{1}H-domain) may also be valuable in the case of highly overcrowded ^{1}H spectra. The resolution in the ^{13}C domain is usually much better and allows individual slices along F2 for each carbon resonance to be extracted and the corresponding ^{1}H sub-spectrum to separately be inspected. The analysis of such slices, i.e. the evaluation of ^{1}H chemical shifts and ^{1}H/^{1}H coupling constants, is improved if the resolution in F2 has previously been enhanced. Transfer the first row of the unprocessed data with the **FID Transmission** command in the **File** pull-down menu into 1D WIN-NMR. Use the sine or sine squared window types and interactively adjust **SSB** to optimize its resolution. Return to 2D WIN-NMR and use this **SSB** value and the choosen window in F2 and a **QSINE** window with **SSB** = 2 in F1. Process the spectrum. After phase correction extract the individual F2 slices for the six carbons, transfer them into 1D WIN-NMR as described above and compare them with the corresponding reference spectrum D:\NMRDAT\GLUCOSE\2D\ 1DREF\GH\001001.1R. Note that spectral analysis is severely hampered, because digital resolution is much worse for the 2D slices compared to the reference spectrum.

Check it in 2D WIN-NMR:

Load the raw data of the magnitude mode ^{1}H/^{1}H- and ^{13}C/^{1}H COSY experiments of peracetylated glucose D:\NMRDATA\FORMAT\DISNMR\GLUCOSE\2D\HH\HHDIS\001001.SER and D:\NMRDATA\FORMAT\DISNMR\GLUCOSE\2D\CH\CHDIS\001001.SER originally acquired in the DISNMR format, but converted into the WIN-NMR format. Inspect and eventually adjust the processing parameters (see instructions given in chapter 2.7.2.2) and process the data to obtain the best spectral quality.

5.3.3 Addition of a Processing Function: s(t) + f(t)

There are three main processing options based on the addition of a processing or correction function to the FID: DC- or Baseline-Correction, Zero-Filling and Linear Prediction LP.

5.3.3.1 DC-Correction/Baseline-Correction

DC-Correction is applied to compensate for a DC-offset of the FID, i.e. a vertical shift of the FID with respect to the zero-line, which occurs in quadrature detection mode if the two channels are not matched to each other. The effect is most pronounced for very weak samples and manifests itself, after Fourier transformation, as a spike in the centre of the spectrum at the "center" or "carrier" frequency.

With 1D WIN-NMR the compensation of DC offsets is the only baseline correction option for 1D FIDs. It is automatically performed prior to any processing when confirmed in the dialog box that appears on screen. The baseline of the FID is corrected with the zero-line set to the mean value of the last part of the FID.

With 2D WIN-NMR baseline corrections are automatically applied for FIDs in t2 and t1 prior to any processing when the Fourier transformation is initialized with the **xf2**, **xf1**, **xfb** or **xtrf** commands. According to the parameter **BC_mod** set for F2 and F1 separately, either a simple DC correction or more sophisticated algorithms are applied to correct the FID baselines in F2 and F1 (Table 5.2), thereby taking into account the detection mode (single/quadrature).

Table 5.2: Settings for **BC_mod** and corresponding effects

BC_mod:	Effect:
no	no baseline correction applied
single/quad	a constant is subtracted from the FID (DC-correction)
spol/qpol	a polynomial of degree 5 is subtracted from the FID
sfil/qfil	filtering of the FID according to Bax/Marion

The latter two baseline corrections are used to suppress water signals, whereas no baseline correction is sometimes preferred if the FID has not decayed strongly towards the end.

Check it in 1D WIN-NMR:

Load the 1D ^{1}H FID of peracetylated glucose D:\NMRDATA\GLUCOSE\1D\H\GH\001001.FID, increase its vertical scale and inspect its last part on the screen. Notice the deviation of the mean FID from the zero horizontal line; switch back and forth between the real and the imaginary part of the FID to recognize a DC offset between the two parts. From the **Process** pull-down menu choose the **DC Correction** option and apply a DC correction. Inspect the last part of the FID again. Fourier transform the FID with/without a DC correction and inspect the region in the centre of the spectrum, where a spike usually appears without DC correction. This effect is not that marked with this raw data acquired on a modern spectrometer. Note that usually a **DC correction** dialog box appears on screen when you start to process any FID, asking you if a DC correction should automatically be applied or not. Once such a DC correction is applied, this correction is stored automatically and when the FID is recalled later the **DC correction** dialog box will no longer appear when you initialize any processing.

Check it in 2D WIN-NMR:

Experiment with the effect of several modes for DC correction available with 2D data. Load the raw data of the magnitude mode 2D COSY spectrum D:\NMRDATA\GLUCOSE\2D\HH\GHHCO\001001.SER and choose **no**, **quad** and **qpol** as **BC_mod** in F2. Note that usually for F1 no baseline correction is applied in the time-domain, i.e. **BC_mod** (F1) = **no**. Fourier transform the data and store the individual spectra using ascending processing numbers. Compare corresponding rows of these three spectra to inspect the effect of different baseline corrections.

5.3.3.2 Zero Filling

Adding zeros at the end of the FID (Zero Filling) as shown in Fig. 5.15, is applied to increase the number of data points prior to Fourier transformation thereby increasing the digital resolution in the spectrum. Increased digital resolution generally improves the shapes of the individual resonances, since each resonance line is defined by more points. Under certain circumstances zero filling also enhances spectral resolution and allows closely spaced signals, e.g. splits due to a small J coupling, to be resolved. The power of resolution enhancement using zero filling should not be overestimated, since spectral resolution is exclusively determined by the length of the acquisition time and the relaxation time T_2 of the corresponding nucleus. Resolution cannot be artificially driven to the extreme by simply adding more and more zeros, i.e. by applying zero-filling several times.

With 1D WIN-NMR zero filling is accessible, along with other less important options, via the **Zero Filling...** option in the **Process** pull-down menu (Fig. 5.20) or by clicking with the right mouse button on the **Zero Filling!** button in button panel, which both open slightly different dialog boxes.

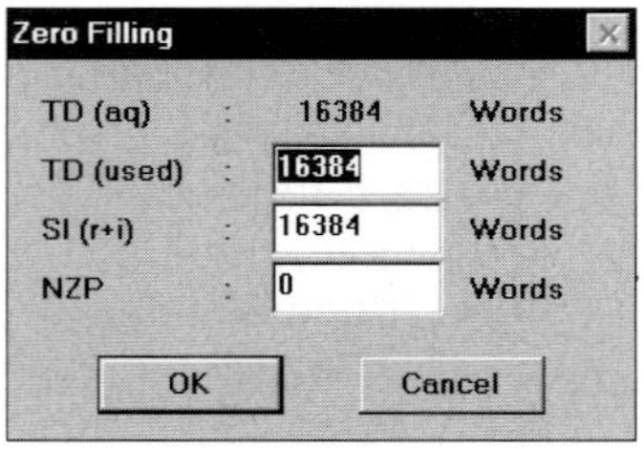

Fig. 5.20: **Zero Filling...** dialog box.

The corresponding parameters in these dialog boxes have been discussed in detail in section 5.2.1. **SI(r+i)** defines the sum of the number of real and the number of imaginary data points that the resulting FID will consist of. With this parameter the FID may be lengthened (zero filling) and also truncated to a specified length. Within the same dialog box the parameters **TD(used)** and **NZP** serve to either cut off the "noisy" final data points of an FID when recorded with an acquisition time of unsuitable length, or to cut off the very first data points of an FID, distorted by some spectrometer perturbations (e.g. pulse breakthrough). Zero filling is initialized either with the command available within the dialog box (**OK** or **Execute**) or directly by clicking with the left mouse button on the **Zero Filling!** button.

With 2D WIN-NMR zero filling is defined simply by setting SI for the F2 and F1 dimension in the **Parameters** dialog box opened with the **General parameters setup** command in the **Process** pull-down menu prior to Fourier transformation.

Check it in 1D WIN-NMR:

Load the proton-FID of peracetylated glucose D:\NMRDATA\GLUCOSE\1D\H\GH\001001.FID. Switch to the **Pts** mode and expand the FID to recognize its "digital" character, i.e. to see the individual data points. Note that only every second data point is connected with a straight line for the real and the imaginary part respectively, which means that in this case 32K real and 32K imaginary points have been acquired. Perform a series of zero-fillings, using the **Zero Filling...** command in the **Process** pull-down menu, with **SI(r+i)** set to 64K (no zero filling), 128K (1x zero filling) and 256K (2x zero filling). Also set **SI(r+i)** to 32K and 16K. Fourier transform each FID, inspect the number of total data points (SI/2 for the real and SI/2 for the imaginary part), phase correct the spectra and store them with increasing processing numbers. Compare the results using the multiple display option. For this purpose select a region with closely spaced signals, i.e. the region of the methyl signals and compare the line shape and the spectral resolution as a function of the number of zero-fillings.

Try out the effect of zero-filling on the 1D ^{13}C raw data of peracetylated glucose D:\NMRDATA\GLUCOSE\1D\C\GC\001001.FID.

Check it in 2D WIN-NMR:

Load the raw data of the gradient enhanced, double quantum filtered 2D COSY experiment D:\NMRDATA\GLUCOSE\2D\HH\GHHCODF\002001.SER. Use equal **SI** values in F2 and F1 and perform three 2D Fourier transformations with **SI** set to 256, 512 and 1K words. Select a suitable row and column, store them separately and compare the corresponding rows and columns obtained with different zero-filling settings using the multiple display option available with 1D WIN-NMR.

5.3.3.3 Linear Prediction

Among the various processing options available to improve the quality of FIDs and the corresponding spectra, Linear Prediction (LP) and the Maximum Entropy Method (MEM) – not available with WIN-NMR – are probably the most exciting and powerful, even though they are not widely used.

LP is mainly used to repair distorted parts of an FID caused either by mis-set acquisition parameters, or introduced by some spectrometer perturbation. These unwanted effects are removed from the corresponding spectra using LP by firstly analysing the non-truncated part of the FID and then predicting and reconstructing the truncated part of the FID. Zeroing the first points of a distorted FID or adding additional zeros at the end of an FID (Zero Filling), as outlined above, represent the most primitive approach to LP, since zeros are "predicted" for the FID region to be "repaired" and the FID region to be extended respectively. However with LP and MEM a more sophisticated approach for prediction is available. LP and MEM are mainly used with data from multidimensional NMR experiments, although a few applications using 1D data have also been proposed. With multidimensional data, LP and MEM can in principle be applied to any and all dimensions, e.g. for 2D data sets to t1 and/or t2.

LP methods are applied for different purposes:

Backward Linear Prediction

Backward LP (Fig. 5.21) is usually applied to "repair" the first few points of an FID, distorted by some spectrometer perturbation or a mis-set acquisition parameter, e.g. incorrect receiver gain. Backward LP is also used to reconstruct an FID back to t=0 in those cases where the start of data acquisition has been delayed, e.g. to exclude unwanted spectrometer noise such as the signals from acoustic ringing, and the first few data points are missing. In this case backward LP cancels or at least suppresses the corresponding spectral artefacts such as baseline roll etc.

If the number of dummy scans in a 2D experiments is set too low, data acquisition may be started before a steady-state e.g. thermal equilibrium, has been established. Consequently the first few increments (data points in t1) may be distorted. With backward LP these "wrong" data points in t1 can be recalculated and replaced using the information gained from the residual increments.

Forward Prediction:

Forward LP (Fig. 5.22) is applied to "complete" non-decayed 1D FIDs, an effect which occurs when the acquisition time has been chosen too short, which would otherwise give rise to baseline distorsions in the final spectrum. Forward LP is also used to replace the last part of a very noisy 1D FID by its predicted counterpart thereby improving the signal-to-noise ratio without sacrifying spectral resolution. Note however that for 1D raw data with a high signal-to-noise ratio and with FIDs decaying close to zero, a situation most common for ^{1}H-FIDs (see Fig. 5.12), the application of forward LP will not improve the spectral quality.

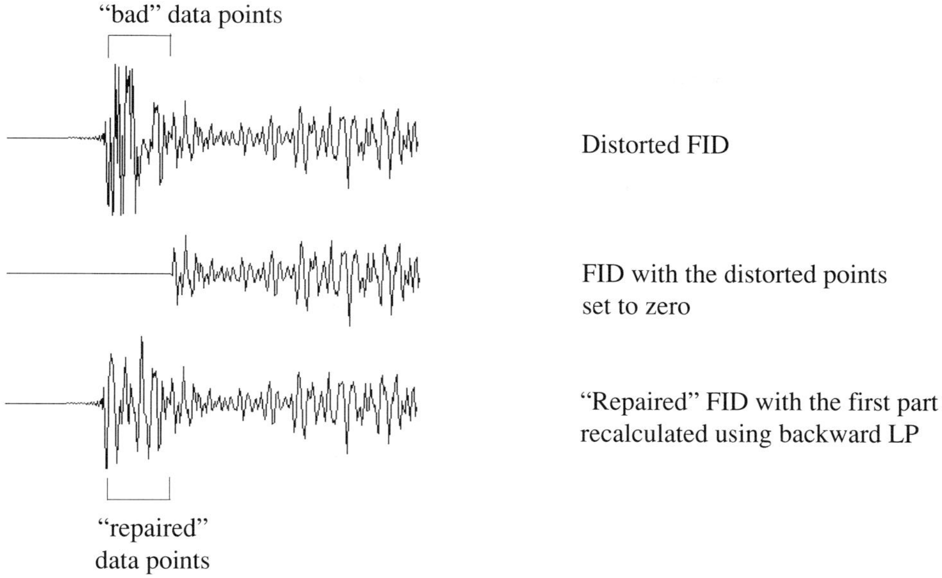

Fig. 5.21: Backward Linear Prediction.

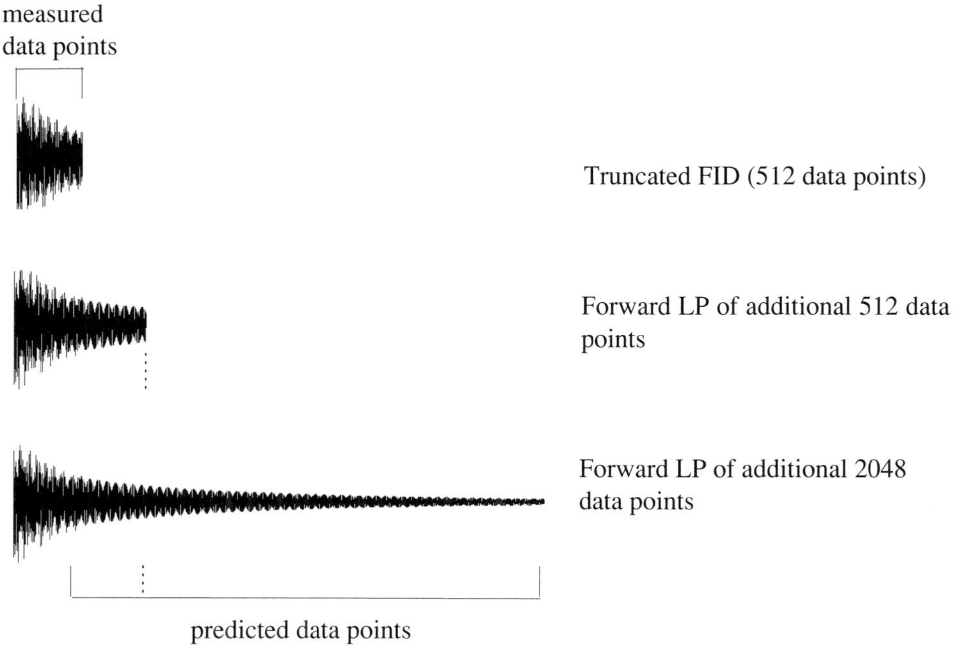

Fig. 5.22: Forward Linear Prediction.

With 2D data sets non-decayed FIDs in both time domains (Fig. 5.13) are very common and a simple Fourier transformation would give rise to truncation effects in the final spectrum. To circumvent such unwanted effects 2D FIDs are usually rigorously damped, using stringent weighting functions to "smoothly" bring the last part of the FID close to zero. However this simple procedure severely impairs spectral resolution and should be replaced by LP, followed by suitable weighting, which both improves spectral resolution and excludes any truncation effects.

Using reasonable amounts of sample and combining 2D experiments with LP in t1 can shorten drastically the total measuring time, since the many increments usually needed for adequate resolution in F1 may be omitted and the missing data points in t1 may be predicted from a few measured increments. The dream to improve simultaneously the signal-to-noise ratio and the resolution in your 2D spectra seems to become a reality!

Forward and backward LP is accessible with 1D WIN-NMR via the **Linear Prediction...** command in the **Process** pull-down menu which opens a corresponding dialog box containing a number of edit fields (Fig. 5.23). **TD(used)** defines the upper limit for the number of measured data points to be analysed and to be used for prediction, whereas **SI(r+i)** is the sum of the number of real and imaginary data points that the predicted FID will contain. Both parameters must be defined via the **Zero Filling...** dialog box prior to starting LP. The meaning of the other parameters together with the corresponding parameters for 2D LP is outlined below.

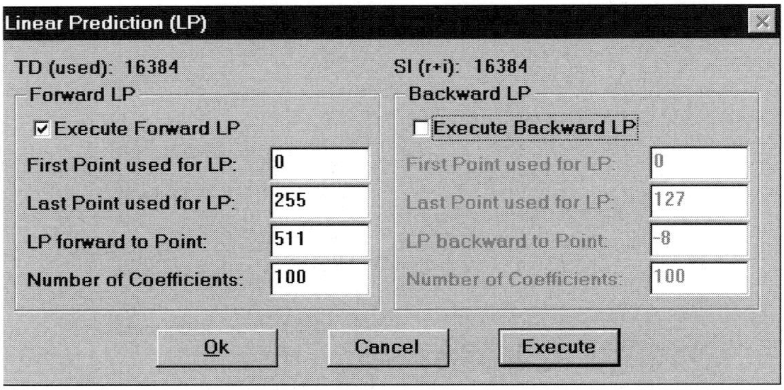

Fig. 5.23: Dialog box for Linear Prediction with 1D WIN-NMR.

With 2D WIN-NMR forward and backward LP is accessible via the **General parameter setup** command in the **Process** pull-down menu where the parameters **ME_mod**, **SI**, **TD** or **TDeff**, **TDoff**, **NCOEF** and **LPBIN** may be selected and edited (Fig. 5.24).

A forward (**LPfr**, **LPfc**) or a backward (**LPbr**, **LPbc**) LP, applied either to real or complex raw data, may be selected using the **ME_mod** parameter. As long as **xf2**, **xf1** or **xfb** are used for the Fourier transformation, the difference between real and complex raw data will be handled internally by 2D WIN-NMR.

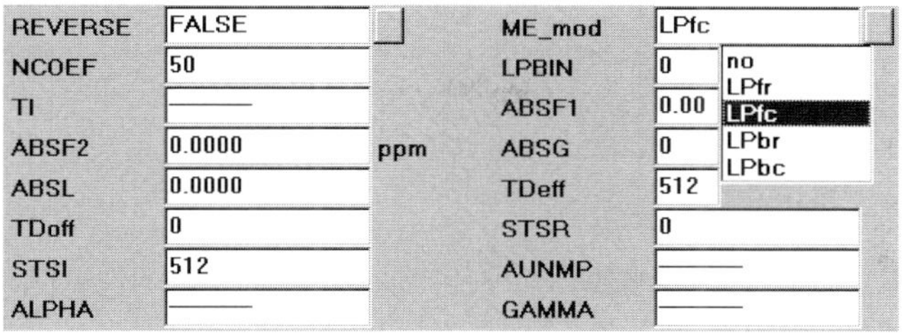

Fig. 5.24: Part of the 2D WIN-NMR Parameters edit box with parameters for Linear Prediction.

The meaning of the parameters used for 1D or 2D forward or backward LP are illustrated below (Figs. 5.25, 5.26). Parameters used for 1D LP are displayed above the FID and parameters used for 2D LP below.

Thus, with 1D forward LP (see Fig. 5.25) the range and the number of measured data points used for prediction is defined by **Last Point used for LP** and **First Point used for LP** respectively (Fig. 5.25). The calculated data points are added to the **TD** (or **TDused**) measured data points up to the limit defined by **LP forward to Point.**

With 2D forward LP (see Fig. 5.25) the number of measured data points used for prediction is **TD**. If **LPBIN** is zero, the predicted data points (**SI** - **TD**) are added to the **TD** measured data points. If **LPBIN** is greater than **TD** but less than **SI** then **LPBIN** defines the upper limit of predicted points and (**LPBIN** - **TD**) data points are calculated and added to the measured **TD** data points (a combination of LP and zero filling). If **TDeff** is greater than zero, then **TD** is replaced by **TDeff**.

With 1D backward LP (see Fig. 5.26) the range and the number of measured data points used for prediction is again defined by the **Last Point used for LP** and the **First Point used for LP** respectively. With **LP backward to Point** positive, the first measured data points (**LP backward to Point** up to **First Point used for LP**) are replaced by predicted points. With **LP backward to Point** negative, **LP backward to Point** data points are added to the beginning of the FID and an equal number of points are discarded from the end of the (zero-filled) FID.

With 2D backward LP (Fig. 5.26) the number and range of measured data points used for prediction is defined by (**TD** - **TDoff**). The number of points contributing to backward LP can be reduced to **LPBIN** if **LPBIN** is set between 0 and **TD**. With **TDoff** positive the first **TDoff** measured data points are replaced by predicted points. With **TDoff** negative, **TDoff** data points are added to the beginning of the FID and an equal number of points are discarded from the end of the (zero-filled) FID. Again, if **TDeff** is greater than zero, then **TD** is replaced by **TDeff.**

Forward Linear Prediction:

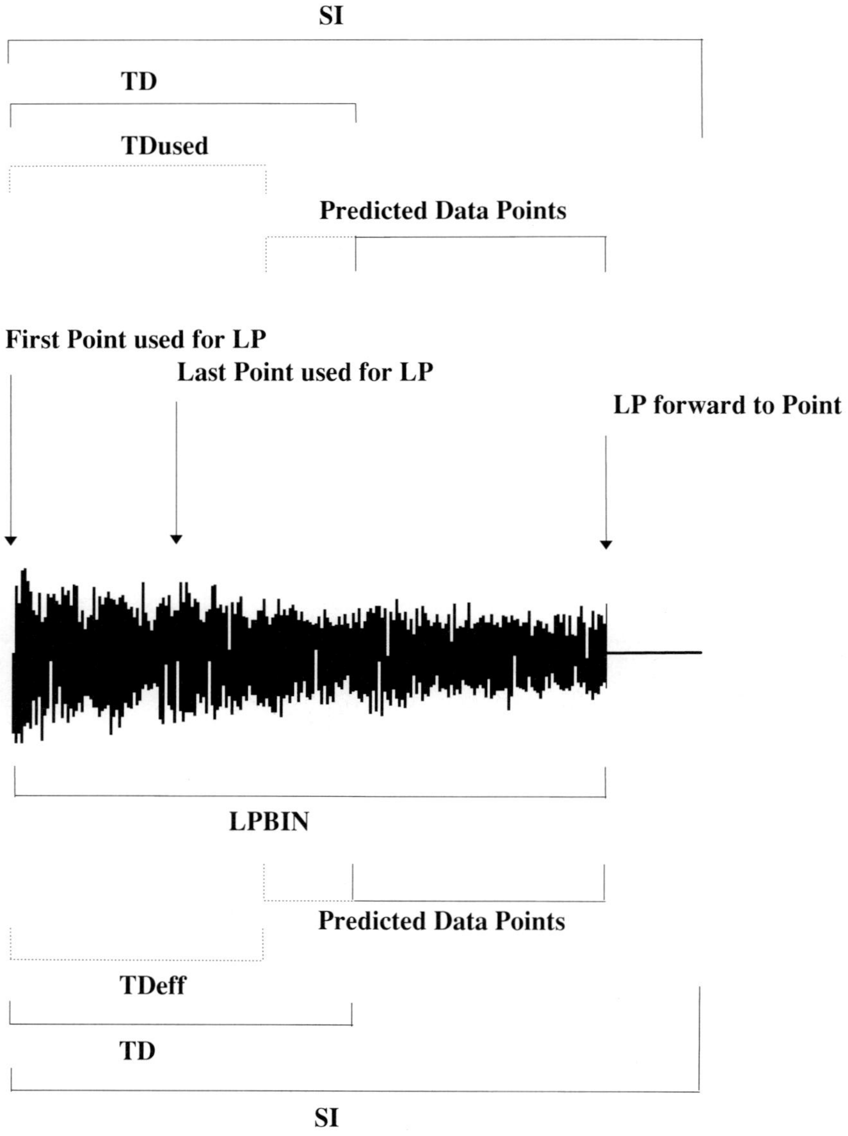

Fig. 5.25: Parameters for a forward Linear Prediction with 1D WIN-NMR (above the FID) and with 2D WIN-NMR (below the FID).

Backward Linear Prediction:

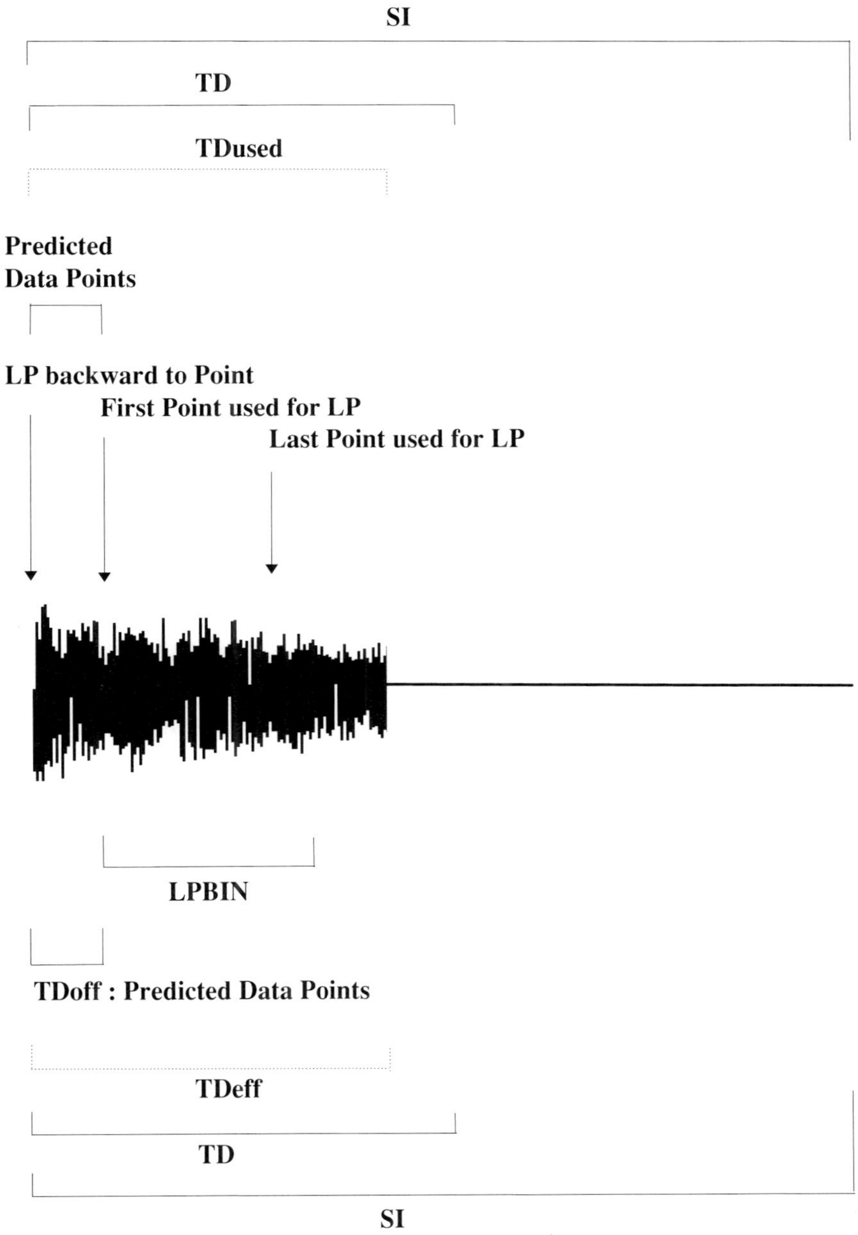

Fig. 5.26: Parameters for a backward Linear Prediction with 1D WIN-NMR (above the FID) and with 2D WIN-NMR (below the FID).

Although theoretical criteria exist for selecting the most appropriate LP parameters, in practice it is sufficient to follow a few empirical rules applicable for both 1D and 2D data sets. The LP parameters may be defined as follows:

Np_{acq}: The number of measured data points used for LP
NP_{pred}: The number of predicted data points
NC: The number of coefficients which depends on the number of spectral peaks **NSP**

The following empirical rules also indicate the range of application and limitations of the LP method:

Rule 1: LP may be applied to any 1D raw data.

Rule 2: LP may be applied along t1 in 2D experiments where no echo evolves in t1 e.g. TOCSY, NOESY, ROESY, HMQC, HMSQ, HMBC. It must not be applied in 2D raw data experiments where an echo may evolve in t1 e.g. COSY and its variants.

Rule 3: In principle, LP may also be applied along t2 in 2D experiments data sets but in practice there are limitations. In inverse experiments all proton resonances evolve in t2 in each serial FID and large values for the number of coefficients must be set (see rule 6) to predict the serial FIDs, giving rise to long calculation times. This problem may be solved rather elegantly by starting from the 2D spectrum, which has been calculated with no LP applied in t2, rather than from the original 2D raw data. Inverse Fourier transformation with respect to F2 yields serial t2-FIDs which no longer include the frequencies of all the proton resonances as did the original data, but only those of the corresponding cross peaks. Consequently, the number of coefficients for LP may be set to a much smaller value. LP in t2 followed by Fourier transformation yields an improved final 2D spectrum.
Since inverse FT is not available with the educational version of 2D WIN-NMR this kind of LP in t2 must be performed on the workstation of the spectrometer, using the XWIN-NMR software, or using the enhanced 32 bit 2D WIN-NMR version.

Rule 4: An upper limit for the number of predicted data points is given by:
$$NP_{pred.} \leq 3 \cdot NP_{acq.}$$

Rule 5: The number of coefficients for simple spectra with well separated peaks is given by:
$$NC = (2 \text{ to } 6) \cdot NSP$$
This condition holds for most 1D ^{13}C spectra and for the rows and columns of simple $^{1}H/^{1}H$ and most heteronuclear 2D spectra.

Rule 6: The number of coefficients for complex spectra with overlapping lines is given by:
$$NC = (1/3 \text{ to } 1/2) \cdot NP_{acq}$$
This condition holds for 1D ^{1}H spectra and for homonuclear and heteronuclear 2D spectra where the rows and/or columns show many signals.

Note that with too many coefficients inherently broad lines may be artificially split up into two lines, noise peaks will be overemphasized and additional lines may appear. Furthermore calculation times increase with the number of coefficients. On the other hand, too few coefficients may have the effect that lines are not well separated, if at all.

Sets of truncated and non-truncated 1D FIDs have been prepared to explore the advantages and limitations of LP in the following *Check its*. To speed up the calculations, the number of time domain data points and the number of resonance lines in the corresponding spectra have been deliberately reduced to a small number. Suitable 2D data sets have also been prepared for use with 2D WIN-NMR (see Table 5.3).

Check it in 1D WIN-NMR:

First load one of the two non-distorted and non-truncated 1D ^{1}H data sets D:\NMRDATA\GLUCOSE\1D\H\GH\013001.FID or ...\015001.FID. Process each data set by simply applying a Fourier transformation, without applying LP, and store the corresponding spectra as reference spectra for use in the subsequent *Check its*.

To become familiar with LP perform a series of forward and backward LPs using these two ideal data sets. In the **Linear Prediction LP** dialog box make sure that the appropriate prediction options (**Execute Forward LP** or **Execute Backward LP**) are enabled. Carefully select the **First Point used for LP** in a forward LP and the point to which a backward LP should be performed (**LP backward to Point**) respectively by expanding the first part of the FID. Note that with digitally filtered raw data the value for this first point to be used for LP is not zero but is close to 124 in this case. Expand the FID to verify this. Using this FID of 2048 data points, perform a series of forward and backward LPs. Following the rules given above choose a different number of points to be used for LP (**First Point used for LP**, **Last Point used for LP**) and to be predicted/calculated respectively (**LP forward to Point/LP backward to Point**). In each case vary the **Number of Coefficients**. Process the linear predicted FIDs the same way and compare the spectra with the spectrum obtained from the original data with no LP applied. Carefully inspect and compare the spectral resolution and the signal shapes obtained with/without LP, using the **Dual** or **Multiple** display of 1D WIN-NMR. Establish your own rules and principles for using LP with your work.

Check it in 1D WIN-NMR:

Load the truncated raw data of the 1D ^{1}H experiment obtained with the receiver gain RG set intentional too high D:\NMRDATA\GLUCOSE\1D\H\GH\014001.FID. Process the data without LP and store the spectrum for comparison with the LP spectra. First compare this spectrum with the spectrum obtained above from the ideal data set D:\NMRDATA\

GLUCOSE\1D\H\GH\013001.FID. Note the baseline artifacts introduced by the truncated FID. In the **Linear Prediction (LP)** dialog box make sure that the **Execute Backward LP** option is enabled and the **Execute Forward LP** option disabled. Set **LP backward to Point** to 124. Following the rules given above vary the residual parameters **First Point used for LP** (recommended: 196), **Last Point used for for LP** (recommended: 2047) and **Number of Coefficients** (recommended: 128 or larger). Carefully inspect the resulting spectra with respect to spectral resolution and signal shapes and compare it with the spectrum obtained without LP.

Hint: Depending on the degree of truncation in your FID, choose the **First Point used for LP** as small as possible

The spectrum obtained with LP using the recommended values is stored as D:\NMRDATA\GLUCOSE\1D\H\GH\014999.1R.

Check it in 1D WIN-NMR:

Load the truncated raw data of the 1D ^{1}H experiment obtained with the acquisition time AQ set intentional too short D:\NMRDATA\GLUCOSE\1D\H\GH\016001.FID. For comparison with the LP improved spectra process the data in two ways and store the results. First simply apply a Fourier transformation to the FID yielding a spectrum with severe truncation effects (sinc-artifacts). Secondly, apply a suitable weighting function (**EM**, **SINE**, **GM**, ...) prior to Fourier transformation to smoothly bring the right hand end of the FID down to zero, thereby suppressing the unwanted truncation effects. Note how this weighting affects the spectral resolution. Compare these spectra also with the ideal spectrum obtained from data SET D:\NMRDATA\GLUCOSE\1D\H\GH\015001.FID.

Prior to any LP calculations perform a Zero Filling with this 1024 point FID by setting **SI(r+i)** = 4096. For forward LP make sure that the **Execute Forward LP** option is enabled and the **Execute Backward LP** option is disabled. Set **First Point used for LP** to 126. Following the rules given above vary the residual parameters **Last Point used for LP** (recommended: 1023), **LP forward to Point** (recommended: 4095) and **Number of Coefficients** (recommended: 256 or larger). Carefully inspect the resulting spectra with respect to spectral resolution and signal shapes and compare it with the spectrum obtained without LP.

The spectrum obtained with LP using the recommended values is stored as D:\NMRDATA\GLUCOSE\1D\H\GH\016999.1R.

Check it in 1D WIN-NMR:

Load one of the ideal 1D ^{13}C data sets D:\NMRDATA\GLUCOSE\1D\C\GC\005001.FID or D:\NMRDATA\GLUCOSE\1D\C\GC\006001.FID showing the region with the signals of the methyl and carbonyl carbons of peracetylated glucose respectively . For comparison with the LP improved spectra process both data sets in two ways and store the results. First, simply apply a Fourier transformation to the 4096 data point FID, yielding a rather noisy spectrum. Secondly apply an exponential weighting function to improve the signal-to-noise ratio. With the acquisition time close to 4 seconds for both data sets set **LB** = 0.25 (matched filter).

For forward LP make sure that the **Execute Forward LP** option is enabled and the **Execute Backward LP** option is disabled. Set **First Point used for LP** to 144 for both data sets. Verify this by expanding the FID. Following the rules given above vary the residual parameters **Last Point used for LP** (recommended: 1023 or even smaller), **LP forward to Point** (recommended: 4095) and **Number of Coefficients** (recommended: 64 or larger). Note that only part of the measured data points (recommended 1023) is used for LP since in this case the residual part of the FID consists mainly of noise. Process the modified FIDs in the same way. Carefully inspect the resulting spectra with respect to signal-to-noise, spectral resolution and signal shapes and compare it with the spectrum obtained without LP.

Note that the signal-to-noise ratio obtained is equal or even better than that obtained using the matched filter and that the spectral resolution is not sacrificed compared to the non-weighted spectrum.

The spectra obtained with LP using the recommended values are stored as D:\NMRDATA\GLUCOSE\1D\C\GC\005999.1R (methyl region) and D:\NMR DATA\GLUCOSE\1D\C\GC\006999.1R (carbonyl region).

Check it in 1D WIN-NMR:

Now load the 64K ^{13}C FID of peracetylated glucose D:\NMRDATA\GLUCOSE\1D\C\GC\001001.FID and apply forward LP to improve the signal-to-noise ratio of the corresponding spectrum. Carefully inspect the FID to define the **First Point used for LP** and **Last Point used for LP** and the **Number of Coefficients**. Follow the rules given before and your experience acquired in the last *Check its*, perform several calculations varying the LP parameters to optimize the spectral quality. Compare the results with the spectrum obtained with/without applying a matched exponential filter and without LP.

A series of 2D HMBC experiments have been performed to demonstrate the benefits of LP with 2D data. Whereas the number of time domain data points in t2 (TD2) remains the same for each experiment, the number of time domain data points in t1 (TD1) and the number of scans (NS) has been varied according to Table 5.3. This allows a comparison of the results obtained with/without the application of LP on the basis of the same measuring times but different number of scans per increment (005001, 006001, 007001) or on the basis of the same number of scans per increment but different measuring times (005001, 008001, 009001).

Table 5.3: Data sets obtained with the HMBC experiments with corresponding parameters

Data file: /GCHICOLR	TD1	NS	TD2
005001	256	4	1024
006001	128	8	1024
007001	64	16	1024
008001	128	4	1024
009001	64	4	1024

Check it in 2D WIN-NMR:

Load the HMBC data file of peracetylated glucose D:\NMRDATA\GLUCOSE\2D\CH\GCHICOLR\005001.FID and calculate the 2D spectrum using appropriate weighting functions (see chapter 5.6). Store the spectrum. Select a few suitable columns and store them as well. This data will serve as a reference and will be compared with the LP improved data. Now load as an example, the data set D:\NMRDATA\GLUCOSE\2D\CH\GCHICOLR\009001.FID which has been acquired using one fourth of the time needed to acquire the previous data set. Using the same weighting functions as above, process the data. Store the spectrum, and the same column numbers as in the previous data set.

Reload the same 2D raw data set, use the same basic processing parameters as above, but now set in the **F1 Processing** parameters list **ME_mod** to **LPfc** and **NCOEF** to 8. Process the data the same way as above, store the 2D spectrum and the same columns numbers. Use the **Multiple** display option in 1D WIN-NMR to compare corresponding columns with respect to signal-to-noise, resolution and spectral artifacts.

The spectrum obtained with LP using the recommended values is stored as D:\NMRDATA\GLUCOSE\2D\CH\GCHICOLR\009999.RR.

Vary **NCOEF** and study the effect of the other parameters (**LPBIN**, **TDeff** and **SI**). Establish your own rules and principles for using LP with your work.

Perform a similar LP study on the remaining data sets.

5.3.4 FID Shift / Adjust Point / Zero Points

The options **FID Shift...** , **Adjust Point** and **Zero Points** can be used to "repair" distorted FIDs but they are of minor importance. Left **FID Shift...** eliminates the distorted first points by simply shifting them out of the FID window and right **FID Shift...** compensates for long preacquisition delays intentionally introduced e.g. to avoid the breakthrough of "spectrometer noise". **Adjust Point** resets a single "wrong" data point in the FID and **Zero Points** replaces the first distorted data points of the FID by zeros. FID shifts can severely affect the phasing of the spectra. A right shift is used in the case of long preacquisition delays where it is not possible to phase the spectrum correctly. A right shift will compensate and allow the spectrum to be phased correctly. A left shift of a FID acquired with the usual preacquisition delay introduces severe phasing problems and is therefore not recommended. Left shifts should be replaced by simply zeroing the first distorted data points, which on the other hand may cause resonance and baseline distortions. Zeroing the first "wrong" data points should therefore be combined with backward LP.

These options are available only with 1D WIN-NMR via the **FID Shift...** , **Adjust Point** and **Zero Filling...** options in the **Process** pull-down menu. Use the **Help** tool for more information about these options, their dedicated panel buttons and how to use them.

Check it in 1D WIN-NMR:

Load a 1D FID and try out the effect of **NZP** (already used in a previous *Check it*), **Adjust Point** and **FID Shift...**. Expand the first part of the FID, switch to a points scale and as a first step set one single point in the first part of the FID to a maximum, store this manipulated FID. Process the data and inspect the effect of this manipulation in the final spectrum. Assume that this distortion has been accidentally introduced by some external perturbation e.g. a "dropped" data point. Test out several ways to repair the damaged FID by readjusting the value of this wrong point, try out several values to approximate this point to an undistorted FID. Load the perturbed FID again and set the first part of the FID, including the "wrong" data point, to zero, by using the **NZP** command in the **Zero Filling...** dialog box. As a third variant perform a left shift of the FID, to shift the "wrong" point outside of the FID window. Apply the same final processing to all the three manipulated FIDs and compare the corresponding spectra, including the non-distorted and original (reference) spectrum. Inspect the baseline and check the phasing behaviour. Perform a backward LP and predict the first part of the FID including the "wrong" data point. Which of all these methods is the best?

5.3.5 Adding two FIDs: $s_1(t) + s_2(t)$

Adding or subtracting different FIDs is commonly used for a number of reasons. For unstable compounds, where the coaddition of blockwise acquired and separately stored 1D FIDs allows the best compromise between signal-to-noise ratio and sample decomposition to be obtained and for spectral editing including:

- Multiplicity selected ^{13}C spectra from DEPT data.
- The calculation of spectra of "pure" compounds from spectra of mixtures.
- The subtraction of FIDs obtained with selective 1D experiments (1D NOE, 1D ROE, ...).

Such calculations could in principle be performed either in the time or the frequency domain. However, it is recommended to do these calculations in the time domain to avoid any loss of spectral quality caused by rounding effects introduced with the Fourier transformation.

Adding or subtracting FIDs is only possible with 1D WIN-NMR and is accomplished with the **File Algebra** option in the **Process** pull-down menu. A dialog box allows you to select the second file and to specify whether FIDs or spectra should be added/subtracted. At the same time the screen is horizontally split into three regions with the original spectrum at the top, the second spectrum in the middle and the result of the arithmetic operation at the bottom. A new button panel appears where several specific options are available (Fig. 5.27). In contrast to the **Dual Display** function, available in the **Display** pull-down menu and discussed in chapter 4, more options exist **with File Algebra**, i.e. to store the result or to move one of the traces with respect to the other. Use the **Help** tool for more information.

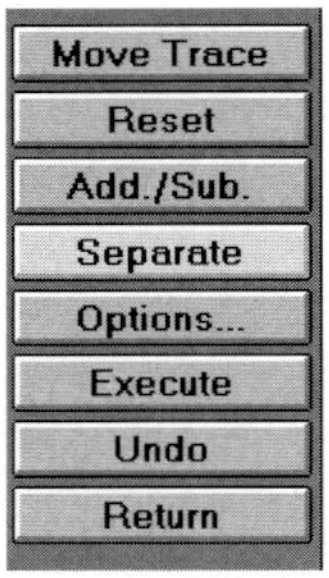

Fig. 5.27: **File Algebra** button panel.

Check it in 1D WIN-NMR:

Load the ^{1}H spectrum of peracetylated glucose D:\NMRDATA\GLUCOSE\1D\GH\001999.1R. From the **Process** pull-down menu choose the **File Algebra** option. In the dialog box select **Spectrum Type** and use the expanded ^{1}H spectrum D:\NMRDATA\GLUCOSE\1D\GH\002999.1R as the second spectrum. Explore the functions of the buttons in the new button panel and try out the effect of the various options. Use the **Help** tool if necessary.

Check it in 1D WIN-NMR:

Load the reference FID of the selectively, homodecoupled ^{1}H experiment D:\NMRDATA\GLUCOSE\1D\GHDC\001001.FID. Use **File Algebra** to calculate the differences between the reference FID and the irradiated FIDs. Store the results and process the data in the same way to obtain the decoupling difference spectra. Use the **Multiple Display** option to display the difference spectra together with the reference spectrum. Note that this same procedure can be applied to 1D NOE or any other selective 1D experiment.

Check it in 1D WIN-NMR:

With modern Bruker spectrometers the selective 1D NOE experiment is usually performed in a "pseudo 2D" mode. The raw data is obtained as a 2D matrix with the individual rows (FIDs) corresponding to the different decoupler frequencies used for the selective perturbation plus one row where the decoupler frequency is set well away from any resonance line (reference FID). Consequently this 2D data matrix must first be decomposed into the individual 1D FIDs before the difference FIDs can be calculated.

From the **File** pull-down menu of 1D WIN-NMR choose the **Filecopy & Convert** option and decompose the original NOE data measured for peracetylated glucose D:\NMRDATA\GLUCOSE\1D\H\GHNO\2D\001001.SER (see chapter 2). Be aware that with this operation the original 2D ser-file will be replaced by a series of 1D files (8 in this case). The original 2D file is no longer available on your harddisk, but may be copied from your CD if necessary.

Load the first decomposed 1D file D:\NMRDATA\GLUCOSE\1D\H\GHNO\2D\001001.FID – note that the extension has changed from SER to FID. From the **Process** pull-down menu choose the **File Algebra** option dialog box and select ...\008001.FID, i.e. the reference FID with the decoupler set off resonance, as the second file. These two FIDs will appear in the top and middle fields of the screen and the result will be shown in the bottom field. The button panel has changed to allow manipulation of the data if necessary. Use the **Help** tool for further information. Click the **Execute** button to initialize the subtraction which will move the result into the top field; click the **Return** button. Store the difference FID using another processing number (e.g. ...\001002.FID). Calculate the other difference FIDs in the same way.

Load the reference FID (...\008001.FID) and calculate the spectrum (weighting with a strong window function, Fourier transformation, phasing). Use the same processing parameters to process the difference FIDs. Exploit the corresponding buttons in the button panel for fast processing. Use the multiple display to inspect the whole series of difference spectra.

Check it in 1D WIN-NMR:

Use the three ^{13}C DEPT data files to calculate multiplicity selected subspectra ("spectral editing"). Load the individual data sets D:\NMRDATA\GLUCOSE\1D\C\GCDP\001001.FID to ...\003001.FID and perform add/subtract operations according to the scheme given below using the functions available with **File Algebra**. Add or subtract files (set y-axis to absolute for this purpose) and determine how to set different factors for the two files to be added/subtracted. Store both the intermediate and the final results using different processing numbers.

With:

D:\NMRDATA\GLUCOSE\1D\C\GCDP\001001.FID	(DEPT-45)	**A**
D:\NMRDATA\GLUCOSE\1D\C\GCDP\002001.FID	(DEPT-90)	**B**
D:\NMRDATA\GLUCOSE\1D\C\GCDP\003001.FID	(DEPT-135)	**C**

calculate:

B	⇨	**CH - Spectrum**
A – C	⇨	**CH_2 - Spectrum**
(A + C) – 1.414 B	⇨	**CH_3 - Spectrum**

Load the DEPT-135 data (data **C**), process the data (weighting, Fourier transformation, phasing) and use the same processing parameters to process the three edited spectra (CH, CH_2, CH_3). Use the multiple display to inspect and compare the three spectra.

5.4 Advanced Processing in the Frequency Domain

After processing in the time domain, Fourier transformation, phasing and basic processing (calibration, peak picking, integration) ahs been performed, additional processing steps to improve spectral quality are at your disposal. This includes operations common to both 1D and 2D spectra e.g. baseline correction in the frequency domain, as well as operations specific to these different types of data sets.

5.4.1 Baseline Correction

Non ideal baselines, deviating from horizontal flat lines, are introduced either by external perturbations or spectrometer imperfections during data acquisition or are caused by mis-set acquisition parameters, e.g. wrong receiver gain. They are most pronounced with weak samples and give rise to problems with the integration of both 1D and 2D spectra and may also affect the appearance of 2D spectra in general. Baseline corrections applied at this stage may overcome such problems.

For 1D WIN-NMR one of three different baseline correction modes (**Offset Correction**, **by defined Points**, **Automatic**) can be selected in the **Baseline Correction Options** dialog box which is accessible via the **Baseline Correction** option in the **Process** pull-down menu (Fig. 5.28) or by clicking with your left mouse button on the **Baseline!** button in the button panel. This **Baseline!** button is most convenient to process series of spectra in the same way. Clicking with your right mouse button on this button opens a **Baseline Correction** dialog box for inspecting and eventually adjusting the baseline function, the maximum number of iterations and for starting the baseline correction.

After selecting the appropriate correction mode, edit fields to set the corresponding correction parameters become active and the functions in the buttons panel are adapted accordingly. Use the **Help** tool for more information.

Offset Correction simply translates either the complete spectrum or a defined interval along the y - dimension.
By defined Points lets you mark certain data points as being on the baseline. From these points a baseline is then calculated according to the specified baseline function, which may (and should) be inspected before the correction is applied.
Automatic performs an automatic baseline correction.
The baseline correction may either be directly applied by clicking the **Execute** button or may first be inspected by clicking the **OK** button, which changes the button panel to *Baseline* mode and program control transfers to different moduls depending on the baseline mode selected. Use the **Help** tool for more informations.

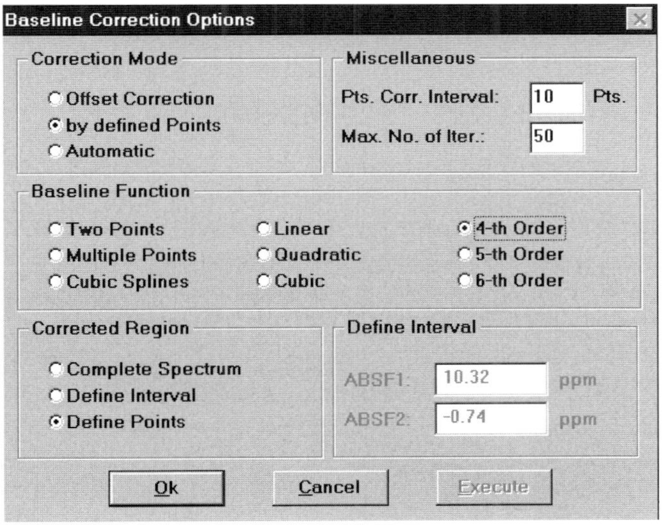

Fig. 5.28: **Baseline Correction** Options dialog box with 1D WIN-NMR.

Check it in 1D WIN-NMR:

Load the proton data of peracetylated glucose D:\NMRDATA\GLUCOSE\1D\H\GH\017001.FID. Fourier transform the FID, phase and store this original spectrum. Increase the vertical scale to inspect the distorted baseline. Try out the three modes of baseline correction using several values for the corresponding parameters by using either the **Baseline Correction** option in the **Process** pull-down menu or by clicking with your right/left mouse button on the **Baseline!** button in the button panel. Store the spectra using ascending processing numbers. Compare the corrected spectra with the original spectrum using the dual display. To illustrate the difference, use the

File Algebra option to subtract the corrected spectra from the original spectrum,

Define integral regions in the original spectrum and use this region file to perform corresponding integrations with the baseline corrected spectra without using any integral corrections (offset, bias) and compare the results.

With 2D spectra an automatic baseline correction may be performed in F2 and/or F1 and may be initialized using the commands available with the **Baseline Correction** item in the **Process** pull-down menu of 2D WIN-NMR (Fig. 5.29). The correction consists of subtracting a polynomial from the spectrum, i.e. from each row and/or column respectively. The processing parameters **ABSF2** and **ABSF1**, accessible in the **General parameters setup** dialog box determine the left and right limits of the regions to be corrected, respectively. **ABSG** is the polynominal's degree and **ABSL** a noise limit factor. Use the **Help** tool for more informations if necessary.

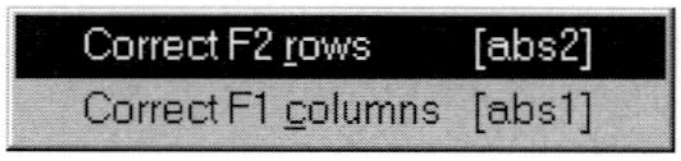

Fig. 5.29: Command box for baseline corrections with 2D WIN-NMR.

Check it in 2D WIN-NMR:

Load the $^{1}H/^{1}H$ COSY data of peracetylated glucose D:\NMRDATA\GLUCOSE\2D\HH\GHHCO\001001.SER, calculate the spectrum and apply a baseline correction in both dimensions. Prior to applying the correction inspect and adjust the **ABSF2** and **ABSF1** to match the spectral regions and set **ABSG** to either 2, 3 or 4. Select suitable rows and columns and compare them with the corresponding slices of the original 2D COSY spectrum.

5.4.2 Additional 1D Specific Processing

5.4.2.1 Deconvolution

Deconvolution is used to disentangle highly crowded spectral regions containing overlapping resonances and to calculate the individual components taking into account different intensities, different linewidths and different lineshape (Lorentzian/Gaussian). Deconvolution is most useful in the case of mixtures of compounds, containing partially overlapping resonance lines and allows the calculation of the ratios of the corresponding components. 1D WIN-NMR offers different possibilities for deconvolution, ranging from a fully automatic to a highly interactive mode. The various modes are available through the **Deconvolution 1** and **Deconvolution 2** options, chosen from the **Analysis** pull-down menu of 1D WIN-NMR. Both options have their own dialog and edit boxes and button panels.

With **Deconvolution 1** you have access to a fully automatic and interactive mode. In the automatic mode only the region used for deconvolution and a few optional parameters (type of lineshape, number of peaks, ...) may be set. Whilst the interactive mode allows you to set the initial values for the parameters controlling the iterative fitting process and to create, edit and delete peaks.

With **Deconvolution 2** you have access to a very flexible manual deconvolution of single lines with the option of mixed lineshape functions (Lorentzian/Gaussian). The measured lineshape of a single line may be approximated manually by setting the signal shape, height, width at half height and position. These options are of some advantage in relaxation analysis.

Use the **Help** routine for for how to use the various deconvolution tools.

Check it in 1D WIN-NMR:

Load the ^{1}H FID of peracetylated glucose D:\NMRDATA\GLUCOSE\1D\H\GH\001001.FID. Zero fill from 64K to 128K, Fourier transform and phase the spectrum. From the **Analysis** pull-down menu choose the **Deconvolution 1** option. Expand the region of the methyl signals (2.15 - 1.95 ppm). Click on the **Region** button and define the region to be used for deconvolution. Perform an automatic deconvolution. Choose 5 peaks to be deconvoluted and check the **Lineshape** type (**Lorentzian**). Inspect the result and try to improve the situation using the options available in the interactive mode. Delete the two automatically created peaks for the two overlapping methyl resonances (**Delete Peaks...**) and then manually create two new peaks (**Create Peaks**). Now choose the region 5.15 - 5.08 ppm and try to deconvolute this part of the spectrum.

Check it in 1D WIN-NMR:

Use the same spectrum and explore the options available with the highly interactive and flexible **Deconvolution 2** option. Use the same two spectral regions as above.

5.4.2.2 Smoothing

Smoothing serves to improve the spectral quality of noisy spectra where the individual lines are distorted by superimposed noise peaks. Its effect is similar to weighting the corresponding FID with a signal-to-noise enhancing function. The stronger the smoothing effect the broader the lines. Smoothing is in some sense the frequency domain counterpart of signal-to-noise enhancement in the time domain. Compared to processing in the time domain, smoothing is less flexible and is mainly used in such cases where the original FID is not available.

Smoothing is accessible in 1D WIN-NMR via the **Process** pull-down menu. A first dialog box is opened and you must specify the number of **Convolution Points** for the smoothing algorithm (Savitzky-Golay). The more convolution points defined the stronger the smoothing effect. Smoothing may be executed directly or may first be inspected and adjusted in an interactive way. The *Spectrum* window is split into two fields with the original spectrum in the upper field and the smoothed spectrum in the lower field. A dedicated button panel appears.

Check it in 1D WIN-NMR:

Load the ^{13}C FID of peracetylated glucose D:\NMRDATA\GLUCOSE\1D\C\GC\001001.FID without applying any weighting function and store is (original spectrum). To study the effect of smoothing enter the smoothing routine and select different numbers of convolution points. Store the results separately and compare them with the original spectrum using the multiple display option. Apply an exponential weighting function to the FID and using several values for the **LB** parameter generate a series of spectra. Compare the smoothed spectra with this series of exponential weighted processed spectra.

5.4.2.3 Derivative

The derivative of a spectrum may be useful for locating the maxima of broad resonance lines and for recognizing non-resolved splittings that manifest themselves as shoulders. Selecting the **Derivative** option in the **Process** pull-down menu of 1D WIN-NMR opens a dialog box where you may specify the type of derivative (**First**, **Second**) to be calculated and the number of **Convolution Points** in the Savitzky-Golay smoothing algorithm to be applied to the derivative. The *Spectrum* window is again split into two fields with the original spectrum in the upper field and the first or second derivative in the lower field. Interactive processing is recommended which allows you to inspect the result and to adjust the number of convolution points prior to execution.

Check it in 1D WIN-NMR:

Load the ^{1}H FID of peracetylated glucose D:\NMRDATA\GLUCOSE\1D\H\GH\001001.FID. Apply a stringent exponential weighting, with **LB** set to 100 Hz to artificially broaden the resonance lines. Phase and store this spectrum. Try out the effect of the first and second derivative. Using the interactive mode vary the number of convolution points.

5.4.2.4 Adjust Point

The **Adjust Point** option available from the **Process** pull-down menu allows you to move single points in y - dimension to correct faulty data points, e.g. spikes. The same function has alreday been used to correct "wrong" values in the FID. **Adjust Point** should not be used to "purify" your sample, by removing the residual signals of impurities or of a solvent.

Check it in 1D WIN-NMR:

Load the ^{1}H spectrum of peracetylated glucose D:\NMRDATA\GLUCOSE\1D\H\GH\001999.1R. Expand the central region and locate the spike at the carrier frequency, which appears in the middle of the spectrum. Use the **Adjust Point** option to "remove" this artifact.

5.4.2.5 Inverse FT

Selecting this option in the pull-down menu **Process** initiates an inverse Fourier transformation. The result of this operation is a FID. Inverse Fourier transformation is of use, if only the spectrum is available and if processing in the time domain (weighting, zero-filling, ..) needs to repeated to improve and optimize the spectrum. Inverse FT is onyl available with 1D WIN-NMR.

Attention: With spectra measured on spectrometers equipped with digital filters (DMX, DRX spectrometers), the automatically performed phase correction (**DMX Phase Corr. ...**) will be applied twice when the newly created FID is Fourier transformed again. This will introduce the baseline roll characteristic for the data of these type of spectrometers. A first order phase correction must then be performed manually by setting the **PHC1** value close to -22000 for the data available in the NMR data base.

Check it in 1D WIN-NMR:

Load the ^{1}H FID of peracetylated glucose in the DISNMR format C:\DEMO1D\DAT\ASP3000\HDIS.00100.FID. Fourier transform the data, phase the spectrum and store it. To simulate bad processing, apply a stringent exponential weighting (**LB** = 4), repeat the Fourier transformation, phasing etc. and store this spectrum as well. Perform an inverse Fourier transformation creating the corresponding FID. Apply appropriate weighting functions (e.g. sine-bell shifted, ...) with the corresponding parameters to correct for the "wrong" initial processing and to optimize the resolution in the final spectrum prior to back transforming this FID.

5.4.3 Additional 2D Specific Processing

In many cases, 2D spectra contain asymmetric artifacts introduced by the experiment. In order to get rid of these unwanted features, some additional processing tools exist and may be performed with 2D WIN-NMR. Corresponding function buttons become accessible via the **Manipulation** command in the **Processing** pull-down menu of 2D WIN-NMR (Fig. 5.30).

Fig. 5.30: Button panel available with the **Manipulation** command in 2D WIN-NMR.

These options should, however, be applied cautiously and the manipulated spectra should be analyzed very carefully, since a loss of spectral information may occur. With one or more of these additional processing steps applied, it is recommended that you plot both the manipulated and the original, non-manipulated spectrum and inspect both spectra when analyzing your 2D data.

5.4.3.1 Symmetrization

Symmetrization is applied to remove any non-symmetrical "artifacts" (ridges, residual peaks not completely removed by phase cycling) from (theoretically) symmetrical data matrices, i.e. $^1H/^1H$ COSY-like spectra, which in many cases helps the subsequent spectral analysis.

2D WIN-NMR offers two symmetrization procedures: **Sym** for magnitude mode spectra and **Syma** for phase sensitive spectra (retaining the original sign). These commands symmetrize the data points on both sides of the spectrum diagonal (which runs from lower left to upper right). Two modes of symmetrization can be selected. Either the larger value of each pair of symmetrical data points is replaced by the smaller or lower value (**lower value**), or both values are replaced by the mean value of the pair (**mean value**).

With J-Resolved spectra a further option for symmetrization is available (**Symj**), which compares the values (y-amplitudes) of pairs of data points situated symmetrically on opposite sides of the horizontal line through the centre of the data matrix. Again the two modes as discussed above can be selected. However in this case, the **Tilt** operation (section 5.4.3.2) must first be applied prior to symmetrization.

Check it in 2D WIN-NMR:

Load the raw data of the 2D magnitude mode COSY experiment of peracetylated glucose D:\NMRDATA\GLUCOSE\2D\HH\GHHCO\001001.SER, calculate the spectrum and increase the vertical scale to see all the artifacts. From the **Process** pull-down menu choose the **Manipulation** option. In the new button panel click on the **Sym** button to symmetrize the spectrum. Perform the symmetrization twice with the symmetrization mode set to **lower value** and **mean value** respectively. Compare the spectral quality and the shape of the symmetrized cross peaks. Repeat this procedure with the data of the gradient enhanced 2D magnitude mode COSY experiment (...\002001.SER). Compare the results with respect to these unwanted artifacts with/without symmetrization.

Check it in 2D WIN-NMR:

Load the raw data of the 2D phase sensitive COSY experiment of peracetylated glucose D:\NMRDATA\GLUCOSE\2D\HH\GHHCODF\001001.SER. Calculate the spectrum and proceed as above, but use the **Syma** command for symmetrization. Compare the results obtained with/without symmetrization.

5.4.3.2 Tilt

After 2D Fourier transformation J-Resolved spectra usually contain a distortion along the horizontal line leading through the centre of the matrix. In order to get rid of this distortion and to separate chemical shifts from homonuclear J-couplings, the whole matrix is tilted. With 2D WIN-NMR a **Tilt** command is available which automatically adjusts the corresponding parameters (Tilt factor) and performs a tilt operation.

Check it in 2D WIN-NMR:

Load the raw data of the 2D J-Resolved experiment of peracetylated glucose D:\NMRDATA\GLUCOSE\ 2D\HH\GHHJR\001001.SER and calculate the 2D spectrum. Inspect the original tilt of the spectrum by calculating and displaying the internal F2 projection, using the **ProjF2** button, accessible via the **Projection** command in the **Display** pull-down menu. From the **Process** pull-down menu choose the **Manipulation** option. Click on the **Tilt** button to perform the tilt operation. Inspect the spectrum and its F2 projection again. Click on the **Symj** button to symmetrize the spectrum.

5.4.3.3 Remove Ridge

Ridges may be introduced into the spectrum either by inappropriate experimental conditions, e.g. if sample spinning with 2D experiments is on during acquisition, or may be introduced by inadequate processing e.g. the "wrong" baseline correction. Ridges are most pronounced at the position of intense singlet peaks and may appear along F1 and/or F2. Whereas the effect of inadequate processing may easily be removed by adjusting the processing parameters, experimentally introduced ridges must be corrected using the **Remove Ridge** function. Other unwanted artifacts such as cross peaks along the F2 axis, either in the middle or at the bottom, may also be removed in this way to "clean" the spectrum. Use the **Help** tool to get more information concerning the **Remove Ridge** option and how to use it

Check it in 2D WIN-NMR:

Load the raw data of the 2D magnitude mode COSY experiment of peracetylated glucose D:\NMRDATA\GLUCOSE\2D\HH\GHHCO\001001. SER and calculate the 2D spectrum. Increase the vertical scale to see all the unwanted peaks and ridges close to the baseline. Note the ridge along F1 on the outer right side of the spectrum and the series of horizontally arranged cross peaks in the center of the spectrum. From the **Process** pull-down menu choose the **Manipulation** option. In the button panel click on the **Rem. Ridge** button to remove both these artifacts. This will open a corresponding dialog box for selecting one of the computation methods, for defining the region to be cleaned and for executing the computation. Inspect the result.

5.4.3.4 Remove Diagonal

Some kind of 2D spectra are characterized by strong diagonal peaks e.g. NOESY or ROESY spectra. The intensity of these diagonal peaks may be many times larger than the intensity of the cross peaks and may make spectral analysis difficult, especially when the sign of the diagonal and the cross peaks is the same (NOESY with large molecules). Removing this diagonal may improve the situation but may also be dangerous, since cross peaks close to the diagonal may be lost in this way.

2D WIN-NMR offers a **Remove Diagonal** function, which allows you to determine the region of the diagonal to be removed in two different ways (numerically or interactively). For more information use the **Help** option.

Check it in 2D WIN-NMR:

Load the raw data of the 2D phase sensitive ROESY experiment of peracetylated glucose D:\NMRDATA\GLUCOSE\2D\HH\GHHRO\001001 .SER. Calculate the phased 2D spectrum and inspect the strong diagonal peaks. From the **Process** pull-down menu choose the **Manipulation** option. In the button panel click on the **Rem. Diag** button to remove the diagonal. Try out the numerical and the interactive way to define the diagonal region. Click on the **Proceed** button in the correspondingly opend dialog box to initialize this manipulation and inspect the result.

5.4.3.5 Remove Peak

If unwanted peaks which have been identified as artifacts are present in your 2D spectrum, they may be removed sequentially using the **Remove Peak** option available with 2D WIN-NMR. This cosmetic operation should be used in situations where either you do not want to symmetrize a spectrum because of the possible loss of information or if symmetrization is not possible as with non-symmetrical 2D data e.g. heteronuclear 2D spectra. For further information use the **Help** tool.

5.4.3.6 Shift/Wrap

2D WIN-NMR offers two additional options allowing you to shift or wrap a 2D spectrum. For further information use the **Help** tool.

5.5 Automatic Processing

5.5.1 Introduction

The automatic processing of either a single file or a series of files would be of great value in standardizing and speeding up the task of processing NMR data. With 1D WIN-NMR, it is possible to automatically process data using most of the options discussed in the previous sections of this chapter. Compared to manual processing the possibilities for optimizing and adjusting processing parameters to the individual data file are limited. Automatic processing starts with loading the 1D FID from disk and ends with plotting the spectrum and storing the fully processed data on disk. The processing jobs, required for processing either a single data file or a series of data files may be set up according to your requirements and may be stored and recalled for later use. To set up these processing job files special dialog boxes for the time and the frequency domain allow you to select which processing steps should be performed and to define the corresponding options and parameters (Fig. 5.31 and 5.32). These options boxes are similar to or are the same as those used with manual processing. You are free to set up a library containing a variety of job files designed to perform specific processing tasks e.g. ^{1}H, ^{13}C, NOE etc.

When automatically processing a series of data files in the same way, the corresponding processing parameters are manually adjusted for one data file representative for the whole series.

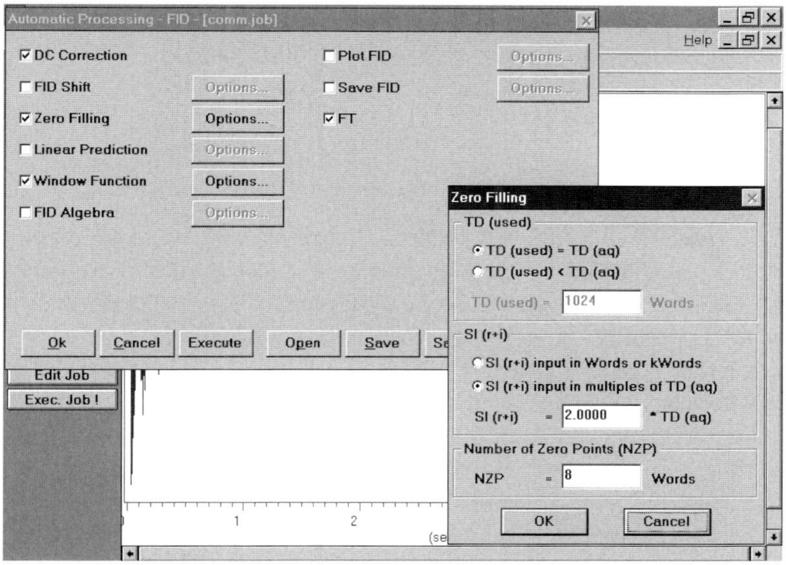

Fig. 5.31: Dialog box to edit the processing jobs in the time domain together with the **Zero Filling** options box.

With 1D WIN-NMR, existing job files are loaded with the **Open Job...** command located in the **File** pull-down menu. Jobs may be edited and inspected using the **Edit Job** button or by clicking with the right mouse button on the **Exec. Job!** button in the button panel which opens the same dialog box. Jobs are started either with the **Execute** command available with this dialog box or may directly be started by clicking with the left mouse button on the **Exec. Job!** button. This last option is most convenient and an alternative to the **Serial Processing** option to process a few or series of data files in the same way.

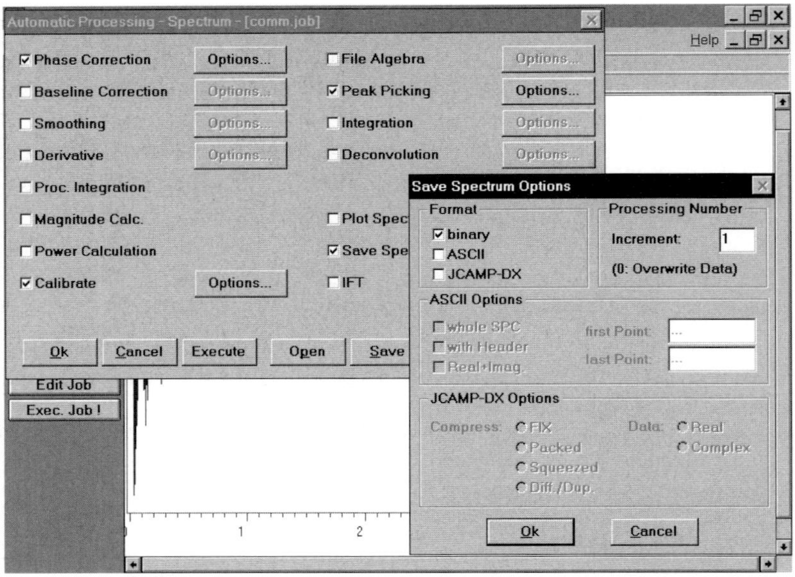

Fig. 5.32: Dialog box to edit the processing jobs in the frequency domain together with the **Save Spectrum Options** dialog box.

For processing a series of data files, e.g. NOE or T_1 data files, the same pre-defined job files are used. To select the data files to be processed a special **Automatic Serial Processing** dialog must be opened with the **Serial Processing** command located in the **File** pull-down menu (Fig. 5.33).

The **Automatic Serial Processing** dialog box allows you to select a series of data files to be processed sequentially. To be processed in this way, the selected data files must all be in the same directory. If data files from different directories are to be serially processed, they must be copied into a common directory prior to being processed.

The same or different jobs may be defined for the selected data files (**use common Job/ use different Jobs**) and each of these jobs may be inspected and edited before the processing of the whole series is started by clicking the **Execute** button. With **the use diff. Jobs** option enabled and clicking the **Define diff. Jobs...** button, an additional **Define Processing Jobs** dialog and edit box is opened containing a list of data files and the corresponding processing job files which may be selected and edited.

The **Automatic Serial Processing** dialog box allows you to select a series of data files to be processed sequentially. To be processed in this way, the selected data files must all be in the same directory. If data files from different directories are to be serially processed, they must be copied into a common directory prior to being processed.

The same or different jobs may be defined for the selected data files (**use common Job/ use different Jobs**) and each of these jobs may be inspected and edited before the processing of the whole series is started by clicking the **Execute** button. With the **Define diff. Jobs...** button enabled, an additional **Define Processing Jobs** dialog and edit box is opened containing a list of data files and the corresponding processing job files which may be selected and edited.

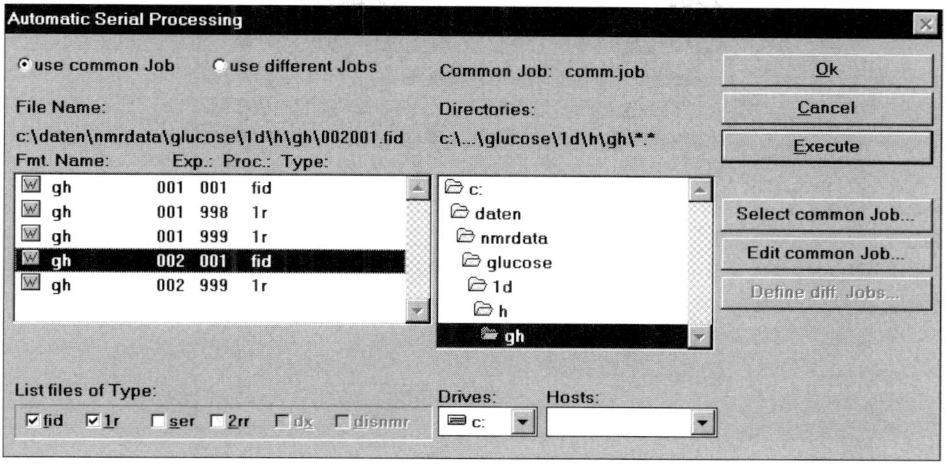

Fig. 5.33: **Automatic Serial Processing** dialog box.

5.5.2 Automatic Processing with Single Files

Check it in 1D WIN-NMR:

Load the ^{1}H FID of peracetylated glucose D:\NMRDATA\GLUCOSE\1D\H\GH\001001.FID and set up a processing job usign the **Edit Job!** button. In the time domain **DC Correction**, **Zero Filling** (**SI(r+i)** =2x**TD(aq)**), **Window Function** (**Exponential**; **LB** = 0.2 Hz) and **FT**. In the frequency domain select **Phase Correction** (**6th Order**), **Save Spectrum** (set **Processing Number Increment** = 1) and **Plot Spectrum** (set the plot parameters according to your preferences). Execute the automatic processing and if you are satisfied with the result, store this job for processing 1D ^{1}H raw data as H.JOB with the **Save** option in the **Automatic Processing** dialog box.

Check it in 1D WIN-NMR:

Load the ^{13}C FID of peracetylated glucose D:\NMRDATA\GLUCOSE\1D\C\GC\001001.FID and set up a processing job. In the time domain select **DC Correction**, **Window Function** (**Exponential**; **LB** = 1.0 Hz) and **FT**. In the frequency domain select **Phase Correction** (**6th Order**), **Peak Picking** (**positive Peaks only**; **X Range**: **whole Spectrum**) of the whole region, **Save Spectrum** (set **Processing Number Increment** = 1) and **Plot Spectrum** (set the plot parameters according to your preferences). Execute the automatic processing and if you are satisfied with the result, store this job for processing 1D ^{13}C raw data as C.JOB.

Modify this job file for processing the ^{13}C DEPT data D:\NMRDATA\GLUCOSE\1D\C\GCDP\003001.FID. Take into account that with DEPT-135 positive and negative peaks appear in the spectrum and set the corresponding processing parameters **Peak Picking** (**pos. and neg. Peaks**) and plot parameters **Plot Spectrum** (**y-Area**: **Whole Spectrum**) accordingly. Execute the automatic processing and if you are satisfied with the result, store this job for processing 1D ^{13}C-DEPT raw data as CDEPT.JOB.

Check it in 1D WIN-NMR:

Load the ^{1}H FID of peracetylated glucose D:\NMRDATA\GLUCOSE\1D\H\GH\002001.FID and set up a processing job. In the time **DC Correction, Window function** (**Exponential**; **LB** = 1.0 Hz) and the **FT**. In the frequency domain include **Phase Correction** (**6th Order**), **Peak Picking** (**positive Peaks only**; **whole Spectrum** in x- and y- direction), **Integration** (**Auto Detect** with the **Sum of Integrals** set to 6), **Save Spectrum** (set **Processing Number Increment** = 1) and **Plot Spectrum** (set the plot parameters according to your preferences). Execute the automatic processing and try out other options and other processing parameters.

5.5.3 Automatic Processing with a Series of Files

Check it in 1D WIN-NMR:

Set up a serial processing job file for the series of ^{13}C FID data files obtained with the Inversion Recovery T_1 experiment D:\NMRDATA\GLUCOSE\1D\C\GCT1\1D\001001.FID to ...\018001.FID. Load the last FID (...\018001.FID) and manually process the data including baseline correction for the FID, exponential weighting with **LB** = 0.5 Hz, FT and phase correction. Note the phase correction values (**PHC0**, **PHC1**). Set up a serial processing job containing all the steps and the same parameters used in the manual processing. In the **Phase Correction Options...** edit box enter the values for **PHC0** and **PHC1** just determined. Also include **Save Spectrum** (set **Processing Number Increment** = 1). Store this job as CT1.JOB. From the **File** pull-down menu choose the **Serial Processing** option which opens the **Automatic Serial Processing** dialog box. Select the T_1 Inversion Recovery files to be processed automatically, make sure that **common job** is activated, check the job settings and parameters if necessary and start the serial processing by clicking the **Execute** button. Inspect the result using the multiple display option.

Check it in 1D WIN-NMR:

Set up a serial processing job file for the series of ^{1}H FID data files obtained with the homonuclear "pseudo 2D" NOE experiment D:\NMRDATA\GLUCOSE\1D\H\GHNO\2D\001001.SER. If not already performed, first decompose the SER file into a series of eight 1D FIDs by choosing the **Filecopy & Convert** option in the **File** pull-down menu. Note that after the **Filecopy & Convert** command the SER file is replaced by the 1D FIDs and is no longer available. Load the reference FID (...\008001.FID) and manually process the data, including baseline correction of the FID, FT and phase correction and store the spectrum. Note the phase correction values (**PHC0**, **PHC1**). Set up a processing job, containing all these steps and the same parameters used with manual processing. In the time domain include **FID Algebra** (set **Factor for Linear Combination with second File** = -1; assign the reference FID file as the second NMR file used for subtraction) (Fig. 5.34) to calculate difference FIDs, the **Save FID** (set the **Processing Number Increment** = 1) for saving this difference FID and the **FT** command.

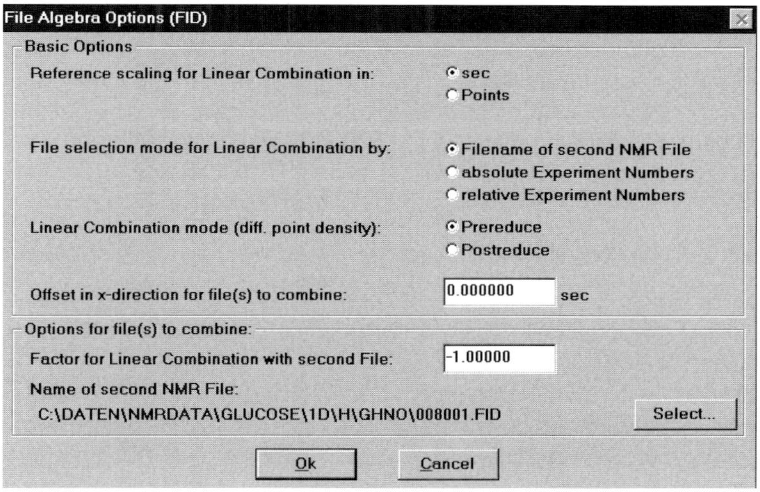

Fig. 5.34: File Algebra Options (FID) dialog box.

In the frequency domain include a **Phase Correction** (using the values for **PHC0** and **PHC1** just determined), **Calibrate** and **Save Spectrum** (again set **Processing Number Increment** = 1) to save the NOE difference spectra. Store this job as HNOE.JOB. From the **File** pull-down menu choose the **Serial Processing** option which opens the **Automatic Serial Processing** dialog box (Fig. 5.33). Select the first 7 NOE data files (...\001001.FID to ...\007001.FID), but explicitly exclude the reference file, for serial processing. Make sure that the **use common job** option is activated, check the job settings and parameters if necessary and start the serial processing by clicking the **Execute** button. Use the multiple display option to inspect the 7 NOE difference spectra, including the reference spectrum as the first or last in the series. If you are not satisfied with the result and if you want to apply a window function to the difference FIDs prior to FT, set up a modified serial processing job file starting, in this case, with the difference FIDs.

Note: When a window function or any other processing in the time domain is to be applied in conjunction with **File Algebra**, then the reference FID must be pre-processed in exactly the same way and must be stored. This pre-processed FID, not the original reference FID, should then be assigned as the second NMR file to be subtracted from all the other FIDs of the series. If this procedure is not carried out, differently processed FIDs will be subtracted from each other giving rise to incorrectly processed difference spectra of unsatisfactory quality.

Note: As an alternative, the manual phase correction of the reference file may be replaced by an automatic phase correction (**Phase Correction**) in the processing job. Each difference spectrum will then be individually phased. As well as being rather time consuming this will give rise to unreliable results. Furthermore, since NOE difference spectra are usually shown with the strong irradiated signals in negative absorption, unless a 180° phase correction is added to the automatically calculated value of **PHC0** these strong signals will appear in positive absorption. To add this 180° phase click on the **Phase Correction Options...** button, in the **Phase Correction** dialog box, set **PHC0** to 180 and select the **Add Numerical Values (PHC0/1)** option.

Check it in 1D WIN-NMR:

Set up a similar serial processing for the series of ^{1}H FID data files obtained with the homonuclear "pseudo 2D" NOE experiment D:\NMRDATA\GLUCOSE\1D\H\GHNO\2D\.001001.SER. In this case, perform the **File Algebra** in the frequency, instead of the time domain. Use the multiple display to inspect the result and to compare it with the results obtained above.

Check it in 1D WIN-NMR:

Set up a serial processing job file for editing the ^{13}C DEPT data, i.e. to automatically calculate three sub-FIDs showing the signals of the CH_3, CH_2 and CH carbons in the corresponding spectra separately. To produce this job file follow the instructions given below. Create a directory e.g. D:\NMRDATA\DEPTEDIT. Copy the original DEPT-45, the DEPT-90 and the DEPT-135 FIDs into this new directory with the file name ...\001001.FID, ...\002001.FID and ...\003001.FID respectively. These original files will not be modified or replaced, but will be used for the calculation of the edited FIDs, i.e. the CH, the CH_2 and the CH_3 sub-FIDs with the file names ...\004001.FID, ...\005001.FID and ...\006001.FID respectively. The serial processing job file is written in a general way such that ...\1001.FID to ...\0003001.FID are always the raw input data files and ...\4001.FID to ...\0006001.FID the DEPT edited files. Consequently, the processing job may be used for any DEPT data set provided that the new data is copied into the correct raw input data files.

D:\NMRDATA\DEPTEDIT\001001.FID	**A**	(DEPT-45)
D:\NMRDATA\DEPTEDIT\002001.FID	**B**	(DEPT-90)
D:\NMRDATA\DEPTEDIT\003001.FID	**C**	(DEPT-135)
D:\NMRDATA\DEPTEDIT\004001.FID	**B**	**CH-Data**
D:\NMRDATA\DEPTEDIT\005001.FID	**A–C**	**CH_2-Data**
D:\NMRDATA\DEPTEDIT\006001.FID	**(A+C) – 1.414B**	**CH_3-Data**

Set up three processing jobs as follows: **Zero filling** (**SI(r+i)** = 1xTDaq, **NZP** = **SI** = 64K to zero the contents of data files ...\4001.FID to ...\0006001.FID), **FID Algebra**, **Save FID** (**Processing Number Increment** =0) in the time domain. Since no FT is applied at this stage, no processing is needed in the frequency domain. Choose the same options for **FID Algebra** as shown in Fig. 5.35, with the **File selection mode for Linear Combination** set to **absolute Experiment Numbers** and with **Factor:** for the last three **Exp. No.** set to **0.000** for all three jobs. Set **Factor:** for calculating the edited data for the first three **Exp. No.** as **0.000/1.000/0.000** (CH-Data, job1), **1.000/0.000/–1.000** (CH_2-Data, job2) and **1.000/–1.414/1.000** (CH_3-Data, job3). Save these three jobs as DEPTCH.JOB, DEPTCH2.JOB and DEPTCH3.JOB respectively.

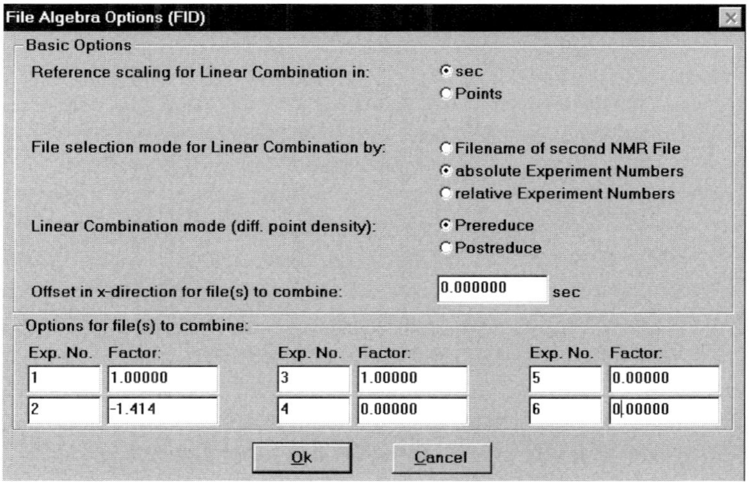

Fig. 5.35: **File Algebra Options (FID)** dialog box to set up the basic options for file(s) to combine (coefficients). Coefficients for calculating the **CH_3-Data**, stored with DEPTCH3.JOB are shown.

From the **File** pull-down menu choose the **Serial Processing** option. Select the files D:\NMRDATA\DEPTEDIT\004001.FID TO ...\006001.FID and choose the **use different Jobs** option. Click the **Define diff. Jobs...** button and in the new dialog box use the **Select Job...** button to assign the correct processing job to each of the data files (DEPTCH.JOB to ...\004001.FID, DEPTCH2.JOB to ...\005001.FID and DEPTCH3.JOB to ...\006001.FID. Start the serial processing with the **Excecute** button. Process the resulting FIDs in the same way according to your preferences (manual/automatic processing; weighting function; baseline correction, peak picking, etc.), inspect the corresponding spectra in a multiple display and plot them.

Note: Alternatively a serial job could be defined with file algebra applied to the spectra rather than the FIDs.

In principle serial processing is not restricted to a set of data acquired with the same experiment as demonstrated above, but may also be applied to data obtained with different types of 1D experiments. Again all the individual data files must first be copied into the same directory (e.g. D:\NMRDATA\SERPROC\) using increasing experimental numbers before processing. The last *Check it* gives an example of how to automatically process a ^{1}H, a ^{13}C and a ^{13}C-DEPT FID. In practise, however, such a "multi-nuclear" automatic processing procedure would be of limited value saving very little time compared to the automatic processing of the individual data files manually initialized one after the other.

Check it in 1D WIN-NMR:

Set up a serial job file to process the standard ^{1}H, standard ^{13}C and three ^{13}C-DEPT FIDs of peracetylated glucose, D:\NMRDATA\GLUCOSE\1D\H\GH\001001.FID, D:\NMRDATA\GLUCOSE\1D\H\GH\001001.FID and D:\NMRDATA\GLUCOSE\1D\H\GH\001001 to ...\003001.FID respectively. Create a subdirectory D:\NMRDATA\SERPROC\ and copy the five files into this directory using increasing experimental numbers:

...\GLUCOSE\1D\H\GH\001001.FID	⇨	...\SERPROC\001001.FID
...\GLUCOSE\1D\C\GC\001001.FID	⇨	...\SERPROC\002001.FID
...\GLUCOSE\1D\C\GCDP\001001.FID	⇨	...\SERPROC\003001.FID
...\GLUCOSE\1D\C\GCDP\002001.FID	⇨	...\SERPROC\004001.FID
...\GLUCOSE\1D\C\GCDP\003001.FID	⇨	...\SERPROC\005001.FID

From the **File** pull-down menu choose the **Serial Processing** option. Select the files D:\NMRDATA\SERPROC\001001.FID to ...\005001.FID and choose the **use different Jobs** option. Click the **Define diff. Jobs...** button and in the new dialog box use the **Select Job...** button to assign the previously saved job files H.JOB, C.JOB and CDEPT.JOB to the correct data files e.g. H.JOB to ...\SERPROC\001001.FID. Start the serial processing with the **Excecute** button and inspect the result. If you are not satisfied with the result adjust the individual jobs to suit your requirements.

5.6 Tables

When importing a data file from a remote computer or directly from the spectrometer the values of the various processing parameters correspond either to some default values or to the settings used by the operator. In the latter case you have simply to perform the same processing operations (weighting, FT, phasing, ...) as applied by the operator with the corresponding parameters stored with the data to get this same spectrum as supplied. Following the philosophy of this book, you should now be competent in processing NMR data and will want to optimize these processing parameters and adjust them to your particular spectroscopic problem. In the following tables, a list of processing parameters is given together with some recommended values for a series of 1D and 2D experiments. Before starting to process any data, check and if necessary modify these processing parameters to suit your own requirements. It is up to you to either accept the recommendations in these tables or to use your own preferred parameter values. The list does not include parameters used for more specialised types of processing such as Linear prediction LP. The parameters associated with these more specialised items have been discussed in the previous sections.

5.6.1 Recommended 1D Processing Parameters

5.6.1.1 ^{1}H Experiments

Table 5.4: 1D ^{1}H Processing Parameters.

Experiment	SI	WDW	Comment
One-Pulse	64K, 128K	NO	
		EM	LB adjusted to t_{aq} and T_2
		SINE, QSINE	SSB: 0, >2 for resolution enhancement or SSB=2 for signal-to-noise enhancement
Selective Decoupling	64K, 128K	EM	LB adjusted to t_{aq} and T_2
		SINE, QSINE	SSB: 0, >2 for resolution enhancement or SSB=2 for signal-to-noise enhancement
1D TOCSY	64K, 128K	EM	LB adjusted to t_{aq} and T_2
		SINE, QSINE	SSB: 0, >2 for resolution enhancement or SSB=2 for signal-to-noise enhancement
1D NOE, 1D ROE	8K, 16K	EM	LB adjusted to improve the signal-to-noise ratio

5.6.1.2 ^{13}C Experiments

Table 5.5: 1D ^{13}C Processing Parameters.

Experiment	SI	WDW	Comment
One-Pulse, DEPT, JMOD (APT)	64K, 128K	EM	LB adjusted to improve the signal-to-noise ratio
T_1 Inversion Recovery	16K, 32K	EM	LB adjusted to improve the signal-to-noise ratio

5.6.2 Recommended 2D Processing Parameters

5.6.2.1 $^1H/^1H$ Experiments

Table 5.6: Processing Parameters for homonuclear $^1H/^1H$ 2D Experiments

F2 Parameters:

Experiment		SI	WDW	SSB	PH_mod	BC_mod
COSY	(magnitude)	512	SINE, QSINE	0	no	no, quad
COSY	(phased)	512, 1K	SINE, QSINE	2	pk	no, quad
TOCSY	(phased)	512, 1K	SINE, QSINE	2	pk	no, quad
NOESY	(phased)	512, 1K	SINE, QSINE	2	pk	no, quad
ROESY	(phased)	512, 1K	SINE, QSINE	2	pk	no, quad
JRES	(magnitude)	512, 1K	SINE, QSINE	0	no	no, quad

F1 Parameters:

Experiment		SI	WDW	SSB	PH_mod	BC_mod	MC2
COSY	(magnitude)	512	SINE, QSINE	0	mc	no	QF
COSY	(phased)	512, 1K	SINE, QSINE	2	pk	no	TPPI
TOCSY	(phased)	512, 1K	SINE, QSINE	2	pk	no	TPPI
NOESY	(phased)	512	SINE, QSINE	2	pk	no	TPPI
ROESY	(phased)	512	SINE, QSINE	2	pk	no	TPPI
JRES	(magnitude)	128, 256	SINE, QSINE	0	mc	no	QF

5.6.2.2 $^{13}C/^{1}H$ Experiments

Table 5.7: Processing Parameters for heteronuclear $^{13}C/^{1}H$ 2D Experiments

F2 Parameters:

Experiment	SI	WDW	SSB	PH_mod	BC_mod	Comment
C/H-COSY (magnitude)	256, 512	QSINE, SINE,	0	no	no, quad	^{13}C-detected
C/H-COSY (phased)	512, 1K	QSINE, SINE	2	pk	no, quad	^{1}H-detected HMQC HSQC HSQC (e/a)
C/H-COSY (magnitude)	512, 1K	QSINE, SINE	0	no	no, quad	^{1}H-detected HMBC

F1 Parameters:

Experiment	SI	WDW	SSB	PH_mod	BC_mod	MC2	Comment
C/H-COSY (magnitude)	256, 512	QSINE, SINE	0	mc	no	QF	^{13}C-detected
C/H-COSY (phased)	256, 512	QSINE, SINE	2	pk	no	TPPI	^{1}H-detetected HMQC, HSQC
						Echo-Antiecho	HSQC (e/a)
C/H-COSY (magnitude)	256, 512	QSINE, SINE	2	mc	no	QF	^{1}H-detected HMBC

5.7 Recommended Reading

Bain, A. D., Burton, I. W., *Quadrature Detection in One and More Dimensions*, Concepts in Magnetic Resonance; An Educational Journal, 1996, *8* (No 3), 191

Derome, A.E., *Modern NMR Techniques for Chemistry Research*, (Chapter 2.5), Pergamon Press, Oxford, 1987

Sanders, J. K. M., Hunter, B. K., *Modern NMR Spectroscopy; A Guide for Chemists*, (Chapter 1.3), Oxford University Press, 1993

6 NMR Data of an Unknown Oligosaccharide

6.1 Introduction

Working through the preceding chapters in this book, you have learnt about the many possibilities for processing NMR data using 1D WIN-NMR and 2D WIN-NMR. You have discovered the advantages and limitations of the various processing options and have learnt how to select and adjust various processing parameters to optimise the final spectrum. The skills you have developed should now be applied in your routine work. But first of all, you can test your understanding of data processing by applying what you have learnt on a more demanding problem, namely the elucidation of the structure of a peracetylated oligosaccharide. The appropriate 1D and 2D data for this compound is contained in the CD-ROM data base. By applying the many processing options, offered by 1D WIN-NMR and 2D WIN-NMR, this unknown structure may be determined. This chapter will demonstrate that with adequate and optimized processing of NMR data a wealth of valuable structural information may and can be obtained from your NMR spectra. Furthermore it will illustrate that additional processing tools designed either to make data analysis more convenient or to assist you in complex cases would be of great use. Such software tools exist and are part of the WIN-NMR software family. Each component of this family is designed to work together in a similar manner to a set of toothed wheels. The application of these other WIN-NMR components is demonstrated in *Modern Spectral Analysis* (volume 3 of this series) and *Intelligent Data Management* (volume 4 of this series).

To assist you in determining the molecular structure of the oligosaccharide a general strategy for unravelling structural problems together with a few practical hints for making data processing and spectra interpretation more efficient are outlined in section 6.2. Readers already familiar and experienced with the strategy of structure elucidation may skip this section. Section 6.3 gives an overview of the NMR experiments applied and the corresponding raw data available for the oligosaccharide. This section also contains tables of results obtained for the peracetylated glucose used in previous *Check its* plus a collection of some typical carbohydrate NMR parameters. By using either the general strategy outlined in this chapter or using your own ideas and preferences, the raw data relating to the unknown peracetylated oligosaccharide should be processed and plotted, each spectrum should be interpreted and the conclusions summarized. Finally, this information should be collated and used to determine the structure. A few *Check its* guide you through the comprehensive NMR data and indicate the parameters and information that may be extracted from individual data sets.

Section 6.4 briefly summarises the information that can be obtained from the various types of NMR experiments and how this information may be used in determining the unknown structure. This section concludes by revealing the structure of the unknown peracetylated oligosaccharide.

6.2 Strategy to Solve Structural Problems

In principle there exists no single strategy for dealing with structural problems. The way a structural problem may be solved depends upon the nature of the problem and the facts already known. Is it sufficient to simply know how the atoms of a molecule are bonded together or is the molecule's stereochemistry important? Should the description include the molecule's conformational behaviour or its dynamic properties? The degree to which a molecules structure should be elucidated determines the number and type of NMR experiment(s) to be applied and makes the whole process of an NMR analysis either more or less time consuming. The strategy involved when there is almost no additional structural information available is very different for that where the structure of a molecule is more or less known and only a few stereochemical problems are left. Typical examples of these extremes are the complete structure elucidation of an unknown natural product and the determination of the relative stereochemistry at two chiral centres of a molecule synthesized in your laboratory. The strategy to be followed also depends on the sample amount, the efficient use of the available spectrometer time, the type of measured NMR data, its quality and the appearance of the corresponding spectra and the availability of either model compounds or reference data.

Furthermore two main strategies exist which are based on spectral analysis and on the use of spectra data bases (see chapter 1). With spectral analysis experiments and data processing are dedicated to extract in subsequent steps the most relevant NMR parameters (chemical shifts, coupling constants and relaxation parameters) and to translate these parameters into structural information. The extraction of NMR parameters is discussed in detail in *Modern Data Analysis* (volume 3 of this series). The second main strategy, described in detail in *Intelligent Data Management* (volume 4 of this series), takes advantage of comprehensive spectra data bases and exploits tools to efficiently compare and predict spectra for unravelling molecular structures.

Following the strategy of spectral analysis at this stage, an NMR investigation could be based on a multiple parameter approach, or may be restricted to the evaluation of one or two NMR parameters only. It is this former multiple parameter approach, together with information obtained from other spectroscopic techniques (MS, IR, UV, ...), which has made high resolution NMR today's most popular tool for structure elucidation.

The steps in a typical multiple parameter NMR analysis (see below), may be subdivided into two groups; the aim of the first group is to assign all the NMR signals in your spectra and, once all these assignments have been established, the aim of the second group is the evaluation the NMR parameters of interest e.g.:

- Chemical shifts.
- Spin-spin connectivities (homo-/heteronuclear) based on scalar coupling over one or more bonds. The determination of coupling constants and the analysis of coupling patterns.
- Spin-spin connectivities (homo-/heteronuclear) based on dipolar coupling and the evaluation of the corresponding NOE (ROE) enhancements.
- Relaxation parameters T_1 and T_2.
- Spin-spin connectivities based on dynamic exchange including the parameters necessary to quantify these processes (k, $\Delta G^{\neq}$, $\Delta H^{\neq}$, $\Delta S^{\neq}$).

For the vast majority of structural problems only a few of these parameters would usually be used and the experiments to measure the more specialised parameters only performed in certain cases. In general though, the more parameters determined and used to solve a complex problem the higher the reliability of the conclusions and the greater the accuracy of the evaluated structure.

There are two final points which should be mentioned at this stage and which concern the analysis of NMR data in general:

1. After having processed your NMR data, a considerable number of plotted spectra may be generated and the sheer amount of paper produced may cause some confusion in the subsequent data analysis. Therefore it is important to be systematic; give each spectrum a clear title, establish – if possible – a structural formula with all the atoms numbered, be careful when assigning peaks in your spectra, cross check each of the assignments, clearly label those peaks which may be assigned unambiguously, in accordance with the numbering in the structural formula and transfer these assignments to all the other spectra as well. The use of standardized layouts for your plotted spectra with the same upper and lower plotting limits, the use of colours to mark signals belonging to the same coupled spin system and the use of lists carrying the NMR parameters for the individual spins is recommended to avoid confusion or even mistakes in your NMR analysis and to facilitate the generation of reports.
2. The accuracy of your measurements should be considered very carefully. The ^{1}H data in the CD-ROM data base was acquired with a digital resolution of 0.5Hz/point or 0.001 ppm/point and the ^{13}C data with a digital resolution of 0.7Hz/point or 0.006ppm/point. For the ^{1}H data in the CD-ROM data base chemical shifts can therefore be quoted to three decimal places e.g. 1.345ppm and ^{13}C data to two decimal places e.g. 134.56ppm, with the proviso that the accuracy of the chemical shifts is ±0.001ppm and ±0.006ppm respectively. However chemical shifts are concentration, solvent and temperature dependent and it is not feasible to use this degree of accuracy when comparing the NMR data with that contained in the literature or a commercial data base. The IUPAC have recommended (see recommended reading at the end of this chapter) that for all NMR data, the solvent, concentration and temperature be included, but in practice these recommendations have been largely ignored. There are no recommendations regarding the accuracy of chemical shifts. Consequently when tabulating your results ensure that where

possible the IUPAC recommendations are followed and that the digital resolution, in both Hertz and ppm, used in recording the spectra is clearly indicated.

6.2.1 General Scheme for an NMR Analysis

The following guideline is restricted to ^{1}H- and ^{13}C NMR data and is based on the series of NMR experiments outlined in chapter 3. The same experiments have been used to obtain the data in the NMR data base. The experimental set up of these popular experiments is relatively straightforward and their combined application has proved to be a very efficient and informative way of solving structural problems. A variety of additional experiments exist and these should be used if and when appropriate in solving special problems.

6.2.1.1 Signal Assignments

1. Process the basic 1D ^{1}H data and find signals representative of a particular type of functional group. Search for characteristic chemical shifts, multiplet structures, signal shapes and check the spectrum for dynamically broadened signals. To confirm your first (tentative) assignments use suitable reference data if available and/or check with standard ^{1}H correlation charts (see recommended reading).
2. Determine the ratio of protons from integrals and establish the (tentative) number of protons if possible. Use suitable signals, e.g. the (singlet) signals of methyl groups, to calibrate your integrals.
3. Try to characterize your sample with respect to its purity; pure, mixture or small amounts of impurities. Duplicates of signals with similar multiplet shapes point to a mixture of isomers.
4. Try to find signals suitable as entering points for subsequent signal assignments.
5. Check the spectrum to establish first J-connectivities. Search for identical J-splittings in the case of simple spectra and/or search for characteristic distortions of line intensities (well known in its simplest form for AB-spectra).
6. Assign signals in your spectra only if an assignment is possible and reliable. Be careful and do not hesitate to label an assignment as tentative if you have any doubts.
7. Designate potential target signals for subsequent selective 1D experiments, if such experiments are planned.
8. Process the data of the basic 1D ^{13}C experiment and determine the number of chemically non-equivalent carbon atoms in your molecule. Find signals representative of a particular type of functional group. To confirm your (tentative) assignments use suitable reference data if available and/or check with standard ^{13}C correlation charts (see recommended reading).
9. Process the 1D ^{13}C DEPT (^{13}C APT) data and establish the carbon multiplicities (CH, CH_2, CH_3). Compare the DEPT spectra with the basic 1D ^{13}C spectrum to assign quaternary carbons (C_q). Make a first cross check with the information obtained from the 1D ^{1}H spectrum.

10. Process the 2D $^{1}H/^{1}H$ COSY or/and 1D $^{1}H\{^{1}H\}$ homonuclear decoupling data. Evaluate the J-coupling network for protons starting from suitable entry points. Assign signals on your spectra if an assignment is possible and reliable. Be careful and indicate tentative assignments. Use the 2D $^{1}H/^{1}H$ TOCSY data if ambiguities arise due to accidental signal overlap (see below).
11. Process the 2D $^{1}H/^{1}H$ TOCSY data and try to recognize subspectra originating from spin systems isolated from each other. Confirm the assignments from the 2D COSY or 1D homonuclear decoupling experiments. Use the TOCSY information to overcome any problems arising from the overlapping of signals in the COSY spectra. Use different colours to mark the components of the different subsystems on the hard copy of your TOCSY and the COSY spectra.
12. Process the 2D $^{1}H/^{13}C$ $^{1}J_{CH}$ correlated COSY data (HMQC, HSQC) and copy all the reliable proton and carbon assignments obtained from the spectra above as well as the carbon multiplicity information (DEPT, JMOD, APT spectra). Cross check the assignments and complete those where either only the proton or the carbon assignment has been obtained. Inspect the cross peaks of the methylene carbons, check whether the attached protons are non-equivalent and if necessary confirm the large geminal coupling in the 2D $^{1}H/^{1}H$ COSY and/or the 1D $^{1}H\{^{1}H\}$ homonuclear decoupling data. Mark the methylene protons in all the previous proton spectra (e.g. with an asterisk). Identify the proton and carbon signals which have not yet been assigned.
13. Process the 2D $^{1}H/^{13}C$ $^{n}J_{CH}$ correlated COSY data (HMBC) and copy all the reliable proton and carbon assignments made for your 2D $^{1}H/^{13}C$ $^{1}J_{CH}$ COSY spectrum. Again include the carbon multiplicity information. Try to assign the quaternary signals and establish long-range interactions between proton bearing carbons and protons on carbon atoms two or more bonds away. Use this information to join together isolated spin systems. Note that with heteronuclear long-range couplings with similar values for $^{2}J_{CH}$ and $^{3}J_{CH}$ ambiguities involving the number of bonds between the coupled spins may arise. These ambiguities may be overcome with the help of the 2D $^{1}H/^{13}C$ HOESY or the 1D $^{13}C\{^{1}H\}$ NOE experiment, which establish heteronuclear through space connectivities. Unfortunately both these experiments are hampered by their inherent low sensitivity.
14. Process the 2D $^{1}H/^{1}H$ NOESY (ROESY) and/or 1D $^{1}H\{^{1}H\}$ NOE (ROE) data and analyse the spectra looking for chemical exchange phenomena and NOEs (ROEs). Use NOE (ROE) connectivities and the corresponding through space interactions to independently confirm geminal and vicinal J-connectivities and to cross check ^{1}H signal assignments that have been based on J-coupling interaction.

    Note: Due to saturation transfer, peaks arising from chemical exchange display different characteristic from true NOEs (ROEs). With chemical exchange the cross peaks in 2D spectra have considerable intensity and are of the samesign as the diagonal peaks while in 1D NOE (ROE) difference spectra, peaks of the same sign and similar intensity as the strong signal of the presaturated spin(s) may be observed. On the other hand signals from NOEs (ROEs) effects are much smaller. They are opposite in sign with respect to the diagonal peaks and the presaturated spin respectively in all ROE experiments and in NOE experiments on small molecules, but are of same sign in NOE experiments on large molecules.

6.2.1.2 NMR Parameter Evaluation

1. Use the 1D ^{1}H and the 2D $^{1}H/^{1}H$ COSY spectrum to evaluate the chemical shifts, the homonuclear coupling network and the corresponding coupling constants. Establish a table including these parameters and the assignments. Omit the δ and J values for those signals that are not first order. If necessary use additional data from 1D $^{1}H\{^{1}H\}$ homonuclear decoupled spectra, 1D ^{1}H TOCSY spectra or 2D phased mode $^{1}H/^{1}H$COSY spectra. Consult *Modern Spectral Analysis*, volume 3 in this series, for more details on analysing second order spin systems and confirming the already evaluated parameters using the WIN-DAISY software tool.
   Note that:
   - when evaluating chemical shifts and coupling constants using 1D $^{1}H\{^{1}H\}$ homonuclear decoupled spectra Bloch-Siegert effects must be taken into account.
   - with 1D TOCSY spectra line intensities deviating from the basic ^{1}H spectrum may occur.
2. If necessary, measure the chemical shifts from the 1D ^{13}C spectrum/1D ^{13}C DEPT (^{13}C APT) spectrum. Draw up a table containing a column for chemical shifts, multiplicities and assignments. Leave enough space in the table to include the T_1 value for each carbon and a section for $^{1}H/^{13}C$ correlations.
3. Examine the 1D $^{1}H\{^{1}H\}$ NOE (ROE) and the 2D $^{1}H/^{1}H$ NOESY (ROESY) spectra, draw up a table including the proton chemical shifts and assignments and indicate among which protons NOEs (ROEs) have been observed. Try to be semi-quantitative in evaluating the NOEs (ROEs) and to label the effects as strong (effects observed among geminal protons), medium or small. When performing a quantitative evaluation follow the general rules, outlined in *Modern Spectral Analysis*, volume 3 in this series.
   at:
   - in contrast to J-coupling the value of the NOEs (ROEs) observed between a given pair of spins H_A, H_B is usually different depending on whether H_A is irradiated and H_B observed or vice-versa and as a consequence the NOE (ROE) data matrix is usually asymmetric.
   - the most useful NOEs (ROEs) for structure elucidation are those observed between spin systems that are spacially close but are in different molecular fragments.
   - in a 1D or 2D NOE (ROE) experiment with the originally perturbed proton spacially close to a proton that is part of a strongly coupled spin system, all the components of that spin system will usually show NOEs (ROEs). This "knock-on" effect should be taken into account when analysing and interpreting the NOEs (ROEs) of strongly coupled spin systems.
4. To the existing table of ^{13}C data add a row showing the ^{1}H chemical shifts and assignments. Inspect the 2D $^{1}H/^{13}C$ $^{n}J_{CH}$ correlated COSY spectrum and in the table

indicate clearly the one-bond and the long-range correlations. Try to evaluate the corresponding cross-peaks as strong, medium or weak. Note that the intensity of the cross-peaks depends upon a number of experimental parameters including the value of delays in the pulse sequence adjusted to the expected $^{n}J_{CH}$ coupling constant.

Pay attention to heteronuclear J-coupling interactions between nuclei of different spin systems, particularly if these spin systems belong to different molecular fragments.

To determine the $^{n}J_{CH}$ heteronuclear coupling constants special 1D and 2D experiments have been developed. For the analysis of 1D ^{1}H coupled ^{13}C spectra using the WIN-DAISY software tool consult *Modern Spectral Analysis*, volume 3 of this series.

5. Determine the T_1 values of the individual carbon nuclei by analysing the data from the ^{13}C Inversion Recovery experiment using the interactive fit routine of 1D WIN-NMR. For further information consult *Modern Spectral Analysis*, volume 3 of this series, and the **Help** tool of 1D WIN-NMR. Add the T_1 values to the ^{13}C NMR data table. Try to rationalise the T_1 values with respect to the evaluated structure and the molecular dynamics of the investigated molecule.

6.3 Processing the NMR Data of an Unknown Oligosaccharide

6.3.1 NMR Data

Figure 6.1 gives an overview of 1D and 2D NMR data available for this peracetylated oligosaccharide. The sample was dissolved in $CDCl_3$ and all the experiments where measured on a 500 MHz Bruker DRX 500 spectrometer.

6.3.2 Reference Data

Tables 6.1 and 6.2 summarize the NMR data obtained for the peracetylated glucose. Use this data as a reference in determining the structure of the unknown oligosaccharide. The data was obtained by direct analysis of the corresponding spectra and with the assistance of WIN-DAISY for the evaluation of ^{1}H chemical shifts and $^{1}H/^{1}H$ J-coupling constants. The ^{13}C T_1 values were evaluated using the T_1 fit routine in 1D WIN-NMR.

oligosacc		
1d		
c		
oc	**^{13}C Basic**	full range, expansion
ocdp	**^{13}C DEPT**	DEPT-45, DEPT-90, DEPT-135
ocjm	**^{13}C JMOD (APT)**	full range
oct1	**^{13}C T_1 Inv. Recovery**	
1d	1D files	variable delay according to vdlist.lst
2d	2D file	original pseudo 2D data
h		
oh	**^{1}H Basic**	full range, expansion
ohhd	**^{1}H {^{1}H} Decoupled**	select. decoupled, different target spins
ohno	**^{1}H {^{1}H} NOE**	
1d	1D files	select. preirradiated, diff. target spins
2d	2D file	original pseudo 2D data
ohro	**^{1}H {^{1}H} ROE**	
freq	1D files	select. perturbed, different target spins
ohto	**^{1}H {^{1}H} TOCSY**	
freq	1D files	select. perturbed, different target spins
tmix	1D files	select. perturbed, incr. mixing time

2d		
1dref	1D Reference Spectra	
oc	**^{13}C Basic**	full range, expansion
oh	**^{1}H Basic**	full range, expansion
ch		
ochco	**^{13}C/^{1}H COSY**	$^{1}J_{CH}$ corr., ^{13}C det., magnitude
ochicolr	**^{13}C/^{1}H HMBC**	$^{n}J_{CH}$ corr., ^{1}H det., magnitude
ochicomq	**^{13}C/^{1}H HMQC**	$^{1}J_{CH}$ corr., ^{1}H det., phased
ochisq	**^{13}C/^{1}H HSQC**	$^{1}J_{CH}$ corr., ^{1}H det., phased
hh		
ohhco	**^{1}H/^{1}H COSY**	magnitude
ohhcodf	**^{1}H/^{1}H COSY**	DQ-filtered, phased
ohhno	**^{1}H/^{1}H NOESY**	phased
ohhro	**^{1}H/^{1}H ROESY**	phased
ohhto	**^{1}H/^{1}H TOCSY**	phased
hj		
ohhjr	**^{1}H/J_{HH} JRES**	magnitude

Fig. 6.1: 1D (top) and 2D (bottom) NMR Data of the peracetylated Oligosaccharide

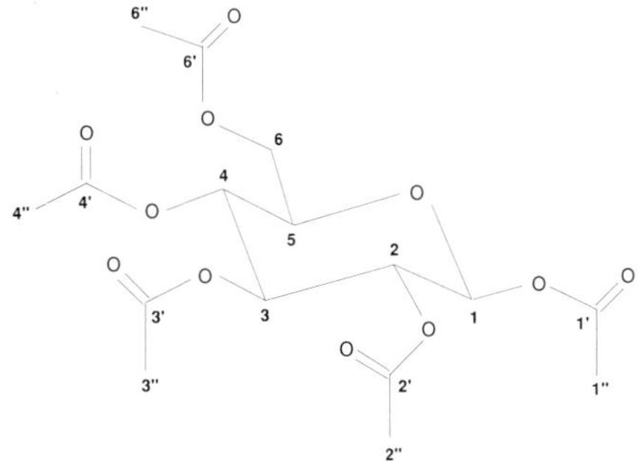

Table 6.1: ^{1}H-chemical shifts, ^{1}H/^{1}H J-coupling constants and ^{1}H/^{1}H NOEs/ROEs of peracetylated β-D-Glucose

H-	δ_H [a] [ppm]	$^{n}J_{HH}$ [a] [Hz]		NOE and ROE [b] [s, m, w] {H-1}	{H-2, H-4}	{H-3}	{H-5}	{H-6a}	{H-6b}
1	5.724	8.3	H-2		w	m	m		
2	5.138	8.3	H-1	w		w			
		9.6	H-3						
3	5. 257	9.6	H-2	m	w [c]		m		
		9.4	H-4						
4	5.133	9.4	H-3			w	w	w	w
		10.1	H-5						
5	3.852	10.1	H-4	m	w	m		m	m
		4.5	H-6a						
		2.2	H-6b						
6a	4.294	4.5	H-5		w		m		s
		-12.5	H-6b						
6b	4.115	2.2	H-5		w		m	s	
		-12.5	H-6a						
1''	2.119	–			not determined				
2''	2.038	–			not determined				
3''	2.017	–			not determined				
4''	2.037	–			not determined				
6''	2.089	–			not determined				

[a] Evaluated from GH\001999.1R, GH\002999.1R with subsequent spectral analysis using WIN-DAISY; the accuracy of the chemical shifts and of the J-coupling constants is ±0.001ppm and ±0.1Hz respectively

[b] Evaluated from GHNO\1D\001nnn; intensity of NOEs/ROEs is: s (strong) 10–20%
m (medium) 5–10%
w (weak) < 5%

[c] NOE/ROE not assignable due to simultaneous saturation of H-2 and H-4

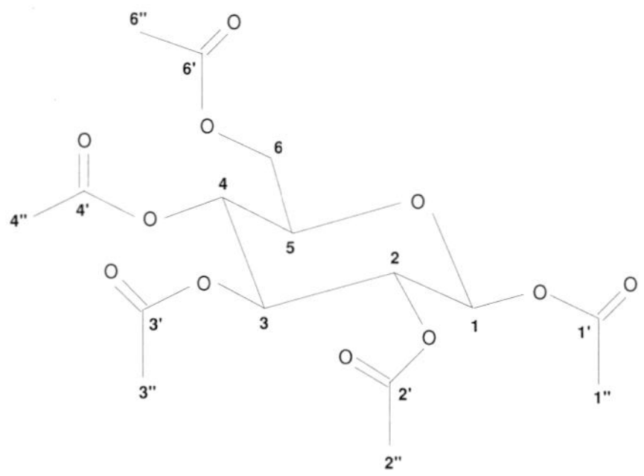

Table 6.2: ^{13}C chemical shifts, ^{13}C T_1 relaxation times and ^{1}H/^{13}C J-connectivities of peracetylated β-D-Glucose

C-	δ_C [a] [ppm]	T_1 [b] [s]	^{1}H/^{13}C ^{n}J-Connectivity with H- [c] [y] 1	2	3	4	5	6a	6b	1"	2"	3"	4"	6"
1	91.72			y			y							
2	70.26		y		y	y								
3	72.81		y	y		y								
4	67.78			y	y		y	y	y					
5	72.74		y		y	y		y	y					
6	61.48					y	y							
1’	168.95	6.1	y							y				
2’	169.24	5.5		y							y			
3’	170.09	5.4			y							y		
4’	169.39	5.7				y							y	
6’	170.59	9.4						y	y					y
1’’	20.81													
2’’	not assigned [d]													
3’’	20.56													
4’’	not assigned [d]													
6’’	20.70													

[a)] Evaluated from GC\001999.1R; the accuracy of the chemical shifts is ±0.006ppm
[b)] Evaluated from GCT1\1D\001nnn.1R; the accuracy of the ^{13}C T1 relaxation times is ±0.1s
[c)] Evaluated from GCHICOLR\001999.RR-004999.RR; y = yes denotes the presence of a cross peak, indicating a heteronuclear long-range coupling
[d)] ^{13}C signals (20.56 and 20.55 ppm) not assignable due to the similar ^{1}H chemical shifts of the corresponding methyl protons and the limited resolution in the ^{1}H domain of the HMQC spectrum (GCHICOMQ\002999.RR)

6.3.3 NMR Data Characteristic of Carbohydrates

^{1}H-chemical shifts R = Carbohydrate

$\Delta\delta$ = **– 0.8** to **– 1.5 ppm for CH**

= **– 0.3** to **– 0.6 ppm for CH_2**

^{1}H/^{1}H J-coupling constants R = Carbohydrate, OAc

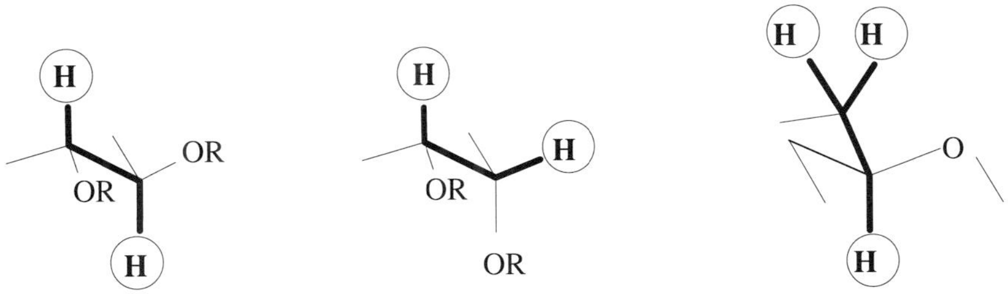

$^3J_{HH}$ = 8-10 Hz **$^3J_{HH}$ = 3-4 Hz** **$^3J_{HH}$ = 3-7 Hz**

Representative ^{1}H/^{1}H NOE/ROE interactions R = Carbohydrate, Oac

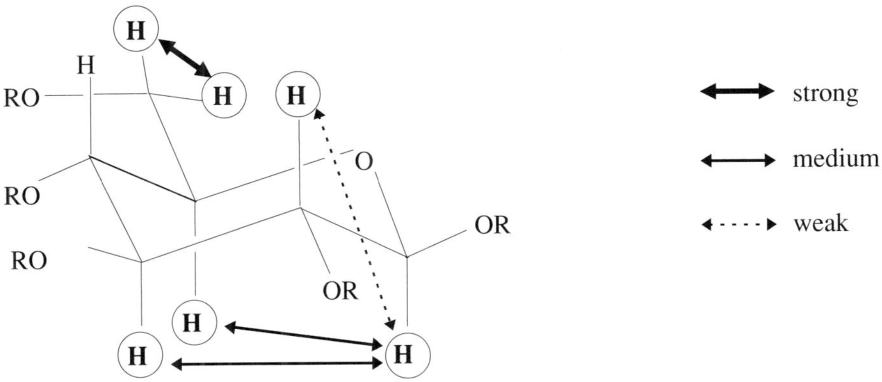

^{13}C-chemical shifts R = Carbohydrate

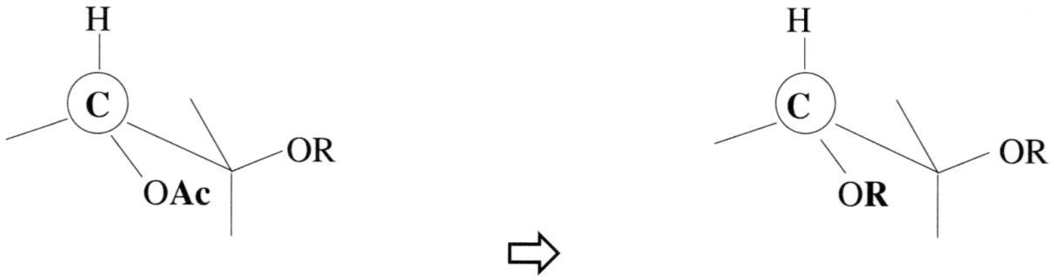

$\Delta\delta$ = 10 to 13 ppm for CH
= 5 to 6 ppm for CH_2

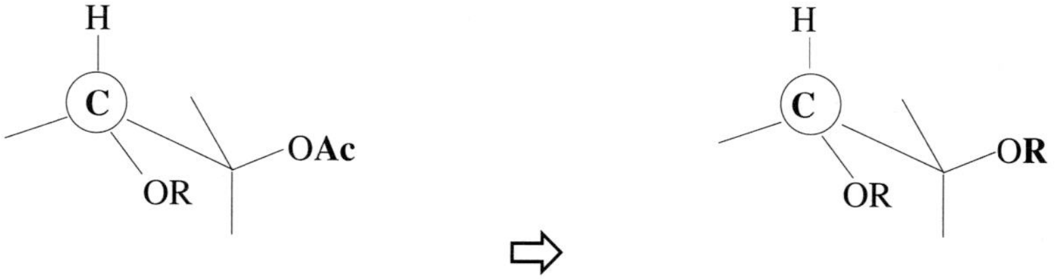

$\Delta\delta$ = 3 to 4 ppm for CH

^{1}H/^{13}C ^{n}J-coupling constants R = Carbohydrate, OAc

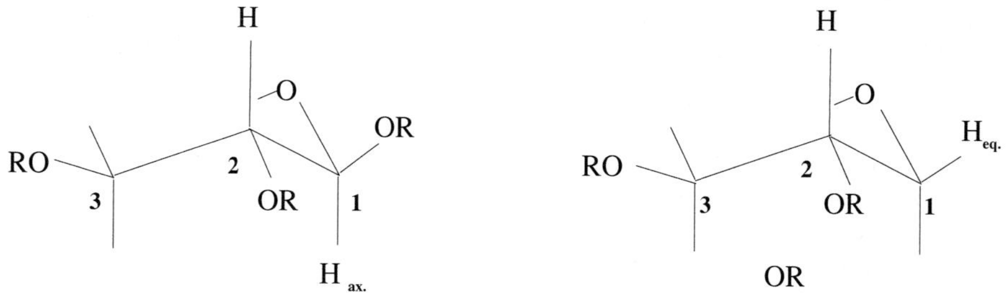

$^{2}J_{CH}$: $^{2}J_{(C\text{-}1,\,H\text{-}2)} = -6$ Hz
$^{2}J_{(C\text{-}2,\,H\text{-}1)} < 1$ Hz

$^{3}J_{CH}$: $^{3}J_{(C\text{-}3,\,H\text{-}1)} = 0$ Hz

$^{2}J_{(C\text{-}1,\,H\text{-}2)} < 1$ Hz
$^{2}J_{(C\text{-}2,\,H\text{-}1)} < 1$ Hz

$^{3}J_{(C\text{-}3,\,H\text{-}1)} = 5 - 6$ Hz

6.3.4 Processing and Analysis of the NMR Data

Check it in 1D WIN-NMR:

Load the raw data of the 1D ^{1}H experiment experiment D:\NMRDATA\OLIGOSAC\1D\H\OH\001001.FID, process it following the procedures outlined in the previous chapters, include calibration, peak picking and integration. If necessary apply suitable window functions to improve the spectral resolution; expand the regions with high signal densities. Plot the spectra according to your own preferences. Try to identify and assign the signals using the reference data of peracetylated β-D-glucose. Search for proton signals which may be suitable as entry points in subsequent spectral analysis e.g. of 2D spectra. Estimate the total number of protons and try to evaluate the number of monomers. Use the second proton spectrum (expansion of the ring-H region) D:\NMRDATA\OLIGOSAC\1D\H\OH\002001.FID as well. Create a list with all your findings concerning the type of protons, the number of methyl groups, monomers, rough chemical shifts, ... and carefully denote those findings which are tentative at this stage.

Check it in 1D WIN-NMR:

Load the raw data of the 1D ^{13}C experiment D:\NMRDATA\OLIGOSAC\1D\C\OC\001001.FID, process it following the procedures outlined in the previous chapters; include calibration and peak picking. If necessary apply a window function to improve the signal-to-noise ratio. Plot the spectrum. Try to identify and tentatively assign peaks using the glucose reference data. Count the number of signals in the various regions (CO, ring-C, methyls) and try to estimate the number of monomers.

Process the ^{13}C DEPT raw data D:\NMRDATA\OLIGOSAC\1D\C\OCDP\001001.FID to 003001.FID and/or the ^{13}C JMOD raw data D:\NMRDATA\OLIGOSAC\1D\C\OCJM\001001.FID as above to determine the multiplicities of the carbon signals. Differentiate between methine and methylene carbons. Use spectral editing to calculate spectra showing the methine and methylene carbon signals separately. Again set up a list with all your findings (chemical shifts, number of C_q, CH, CH_2, CH_3, the number and eventually the type of monomers) and cross-check them with previous findings.

Check it in 2D WIN-NMR:

Load the raw data of one of the 2D $^{1}H/^{1}H$ COSY experiments, e.g. D:\NMRDATA\OLIGOSAC\2D\HH\OCHH\001001.SER, process it following the procedures outlined in the previous chapters. Set up a layout with the individually measured 1D ^{1}H spectra in the directory D:\NMRDATA\OLIGOSAC\2D\1DREF as projections. Expand the regions with high signal densities if necessary and plot the spectra. Use suitable proton signals as entry points and try to evaluate the coupling network. Be aware of the limited digital resolution of 2D spectra and be careful when establishing the coupling network in regions with high signal densities. Use the 2D ^{1}H/^{1}H TOCSY data to circumvent and overcome such problems or use additional data if necessary (see next *Check it*).

Check it in 1D/2D WIN-NMR:

Load the raw data of the 1D ^{1}H TOCSY experiment D:\NMRDATA\OLIGOSAC\1D\H\OHTO\FREQ\001001.SER to 003001.SER, process it following the procedures outlined in previous chapters; include calibration and peak picking. Use the multiple display option to compare the individual spectra. Determine the number of monomers. Compare the spectra of the monomers with your analyzed COSY spectrum and use the TOCSY data to clarify any assignment problems in the COSY spectrum. Use different colours to mark the signals of the individual monomers in both the COSY and the 1D ^{1}H spectra.

Load the 1D TOCSY data acquired with increasing mixing times D:\NMRDATA\OLIGOSAC\1D\H\OHTO\TMIX\001001.FID to 030001.FID, process it and inspect it as a multiple display. Set up a suitable layout and plot the spectra. Confirm the coupling network obtained from the COSY spectra and establish any missing connectivities. Try to identify the anomeric protons and assign the residual signals for all monomers. Copy these assignments to all the previous ^{1}H spectra to simplify further spectral analysis.

Expand the 1D ^{1}H TOCSY spectra of the individual monomers and try to evaluate or at least estimate chemical shifts and coupling constants. Use the data of either the 1D homonuclear decoupling experiment D:\NMRDATA\OLIGOSAC\1D\OHHD\001001.FID to 006001.FID, the 2D phase sensitive COSY experiment D:\NMRDATA\OLIGOSAC\2D\HH\OHHCODF\001001.SER or the 2D J-Resolved experiment D:\NMRDATA\OLIGOSAC\2D\HJ\OHHJR\001001.SER to simplify the evaluation of coupling constants if necessary.

Use the reference data and the characteristic carbohydrate NMR parameters (section 6.3.3) to determine the stereochemistry at the individual ring carbons, and try to determine the connectivity among the individual monomers.

Check it in 1D/2D WIN-NMR:

Load the raw data of the selective "pseudo 2D" ^{1}H {^{1}H} NOE experiment D:\NMRDATA\OLIGOSAC\1D\H\OHNO\2D\001001.SER. Decompose the data matrix into a series of 1D FIDs, calculate the difference FIDs and process the whole series following the procedures as outlined in previous chapters. Exploit the options for automatic and serial processing for this purpose. Use the multiple display option to compare the individual NOE spectra and plot the whole series together with the reference spectrum. Load, process, plot and examine the data of the 2D ROESY D:\NMRDATA\OLIGOSAC\2D\HH\OHHRO\001001.SER and 2D NOESY D:\NMRDATA\OLIGOSAC\2D\H\OHHRO\001001.SER experiments. Inspect the spectra and assign all the protons according to the previous spectra. Then use the internuclear through-space interactions to independently confirm the structures of the monomers deduced from chemical shifts and J-coupling data. Search carefully for NOEs (ROEs) among protons of different monomers (inter-residue NOEs (ROEs)) to determine the connection points and to unequivocally establish the overall structure.

Hint: Use the multiple display option and display each of the NOE (ROE) difference spectra together with the 1D TOCSY spectra of the individual monomer units to recognize and assign intra- and inter-residue NOEs (ROEs) respectively. Use different colours to differentiate between the spin systems of the individual monomers and use a special colour to mark NOEs among different monomers on your hardcopy. Cross check your findings with the conclusions extracted from previous data and carefully check the consistency of your ^{1}H assignments in all the spectra.

Check it in 2D WIN-NMR:

Load the raw data from one of the 2D $^{1}J_{CH}$ COSY experiments (HMQC, HSQC, HSQC-e/a), e.g. D:\NMRDATA\OLIGOSAC\2D\CH\OCHICOMQ\001001.SER, process it following the recommendations given in previous chapters and plot it according to your preferences. Assign the carbon resonances with the proton assignments taken from previous data. Check your tentative assignments in the 1D ^{13}C spectra and if possible complete any missing assignments.

Check it in 2D WIN-NMR:

Load the raw data from one of the 2D $^{n}J_{CH}$ COSY experiments (HMBC), e.g. D:\NMRDATA\OLIGOSAC\2D\CH\OCHICOLR\001001, process it following the recommendations given in previous chapters and plot it according to your preferences. First copy the ^{1}H- and ^{13}C assignments established with the 2D $^{1}J_{CH}$ COSY spectrum. Second mark those cross peaks originating from incompletely suppressed $^{1}J_{CH}$ connectivities then search for long-range connectivities. Interpret the various connectivities on the basis of the oligosaccharide structure and the corresponding reference data and rules given before.

Check it in 1D WIN-NMR:

Calculate the T_1 values for the carbonyl carbons of the oligosaccharide. Load the raw data of the "pseudo 2D" ^{13}C T_1 inversion recovery experiment

D:\NMRDATA\OLIGOSAC\1D\C\OCT1\2D\001001.SER. Decompose the 2D data matrix into a series of 1D FIDs, process and plot them according to the recommendations given in previous chapters. Exploit the options for automatic and serial processing. Determine the T_1 values for the individual carbon nuclei using the interactive fit routine of 1D WIN-NMR. Use the **Help** information if necessary.

6.4 The Structure of the Oligosaccharide

The findings and structural conclusions extracted from the NMR data available for this oligosaccharide may be summarized as follows:

^{1}H-Spectra: OH\001001, OH\002001

1. 3 anomeric proton doublets at 5.708, 4.682 and 4.434 ppm respectively
2. 11 CH_3 singlets of acetoxy groups in the region of 2.15 - 1.98 ppm
3. A few signals of unknown impurities
4. Integration values: 21 ring protons, 33 protons of 11 methyl groups
5. 1 anomeric proton (5.708 ppm) in the range for OAc as the substituent at the anomeric carbon
6. 2 anomeric protons (4.682 and 4.434 ppm) in the range for OR (R = carbohydrate) as the substituent at the anomeric carbon
7. J values for the anomeric protons in the range 7.5 - 8.5 Hz
8. Doublets of the anomeric protons may be suitable as entry points for subsequent spectral analysis or may serve as target spins for additional selective experiments

Conclusions:

- 3 monomers A, B, C
- 2 anomeric carbons connected to another monomer ($C_{anomeric}$–OR)
- 1 anomeric carbon acetylated
- All 3 anomeric protons in axial α-position

^{13}C-Spectra: OC\001001, OC\002001, OCDP\001001-003001, OCJM\00100

1. 11 signals (C_q) of carbonyl carbons in the range of 171 - 169 ppm
2. 3 methine signals in the range of 102 - 91 ppm (range of anomeric C)
3. 12 methine and 3 methylene signals of ring carbons in the range of 78 - 62 ppm
4. 11 methyl signals (2 of them not resolved)
5. A few weak signals of unknown impurities
6. 2 anomeric C (101.26 and 100.14ppm) in the range for OR (R = carbohydrate) as the direct substituent
7. 1 anomeric C (91.93ppm) in the range for OAc as the direct substituent
8. One of the methylene signals characteristically shifted to higher frequency with respect to the others

Conclusions:

- 3 monomers A, B, C
- 2 anomeric carbons connected to another monomer ($C_{anomeric}$–OR)
- 1 anomeric carbon acetylated
- 2 methylene carbons acetylated
- 1 methylene carbon connected to other monomer (CH_2–OR)

2D ^{1}H/^{1}H COSY Spectra: OHHCO\001001-002001, OHHCODF\001001-002001
1D ^{1}H{^{1}H} Homonuclear Decoupling Spectra: OHHD\001001-006001

1. Coupling network evaluated starting with either the anomeric protons (some ambiguities in the range 5.2 - 4.9 ppm) or with H-5 or H-$6_{a,b}$ protons (showing characteristic multiplets) as entry points
2. Assignment of the ring protons for the 3 monomers (see Fig. 6.3) and estimates for the corresponding chemical shifts
3. H-$6_{a,b}$ of monomer A, H-1, H-2 of monomer B and H-1 of monomer C characteristically shifted to lower frequency with respect to the corresponding protons in the other monomers

Conclusions:

- 3 monomers A, B, C
- C-6 of monomer A, C-2 of monomer B and C-1 of monomer C are connected to another monomer, with monomer B as the central unit of this trisaccharide

2D ^{1}H/^{1}H TOCSY Spectrum: OHHTO\001001
1D ^{1}H{^{1}H} TOCSY Spectra: OHTO\FREQ\001001-003001
OHTO\TMIX\001001-030001

1. Proton signal assignments (Fig. 6.3), corresponding chemical shifts and estimated homonuclear J-coupling constants for all 3 carbohydrate monomers
2. H-$6_{a,b}$ of monomer A, H-1 and H-2 of monomer B and H-1 of monomer C characteristically shifted to lower frequency with respect to the corresponding protons in the other monomers
3. $^3J_{HH}$ coupling constants for vicinal couplings among methine protons in the range of 7.5 (anomeric H) - 10Hz

Conclusions:

- 3 monomers A, B, C
- C-6 of monomer A, C-2 of monomer B and C-1 of monomer C are connected to another monomer, with monomer B as the central unit of this trisaccharide
- Mutual *trans* diaxial arrangement of the methine-H throughout, i.e. glucose-monomers

2D $^{1}J_{CH}$ $^{1}H/^{13}C$ COSY Spectra: OCHCO\001001, OCHICOMQ\001001-002001, OCHICOSQ\001001-002001

1. $^{13}C/^{1}H$ correlations of ring carbons and protons, ^{13}C signal assignments (Fig. 6.4) and corresponding ^{13}C chemical shifts
2. C-6 of monomer A, C-1 and C-2 of monomer B and C-1 of monomer C characteristically shifted to higher frequency with respect to the corresponding carbons in the other monomers

Conclusions:

- C-6 of monomer A, C-2 of monomer B and C-1 of monomer C are connected to another monomer, with monomer B as the central unit of this trisaccharide

2D $^{n}J_{CH}$ $^{1}H/^{13}C$ COSY Spectra: OCHICOLR\001001-001004

1. $^{13}C/^{1}H$ correlations through 2 and 3 bonds allows a cross check of ^{1}H and ^{13}C signal assignments for ring carbons and protons (Figs. 6.3, 6.4)
2. $^{13}C/^{1}H$ correlations through 3 bonds assigns all carbonyl carbons and $^{13}C/^{1}H$ connectivities through 2 bonds assigns the methyl proton signals – $^{1}J_{CH}$ connectivities assigns the methyl carbon signals
3. $^{13}C/^{1}H$ long-range correlations between ^{1}H and ^{13}C nuclei of different monomer units (e.g. between C-2 of monomer B and H-1 of monomer C, between H-2 of monomer B and C-1 of monomer C, between C-6 of monomer A and H-1 of monomer B and between C-1 of monomer B and H-$6_{a,b}$ of monomer A) checks the connectivity between different monomer units

Conclusions:

- Monomer A is connected to monomer B (C-6A–O–C-1B)
- Monomer C is connected to monomer B (C-1C–O–C-2B)

1D $^{1}H\{^{1}H\}$ NOE Spectra: OHNO\1D\001001-014001
1D $^{1}H\{^{1}H\}$ ROE Spectra: OHRO\FREQ\001001-009001
2D $^{1}H/^{1}H$ NOESY Spectrum: OHHNO\001001
2D $^{1}H/^{1}H$ ROESY Spectrum: OHHRO\001001

1. Evaluation of network of dipolar couplings , i.e. spatial proximity among protons
2. All three monomers show intra-residue NOEs/ROEs among H-1, H-3, H-5 and among H-2, H-4, H-$6_{a,b}$ respectively
3. Inter-residue NOEs/ROEs are visible among H-1 of unit C and H-2 of unit B and among H-1 of unit B and H-6a (and eventually H-6b) of unit A.

Conclusions:

- Trisaccharide is built up of three β-D glucose monomer units (A, B, C)
- C-6 of unit A is "connected" to C-1 of unit B (C-6A–O–C-1B)
- C-2 of unit B is "connected" to C-1 of unit C (C-2B–O–C-1C)

1D ^{13}C T_1 Spectra: OCT1\1D\001001-018001

1. The measured ^{13}C T_1 values are largest for the two carbonyl carbons of the acetyl groups attached at C-6 of monomer units B and C

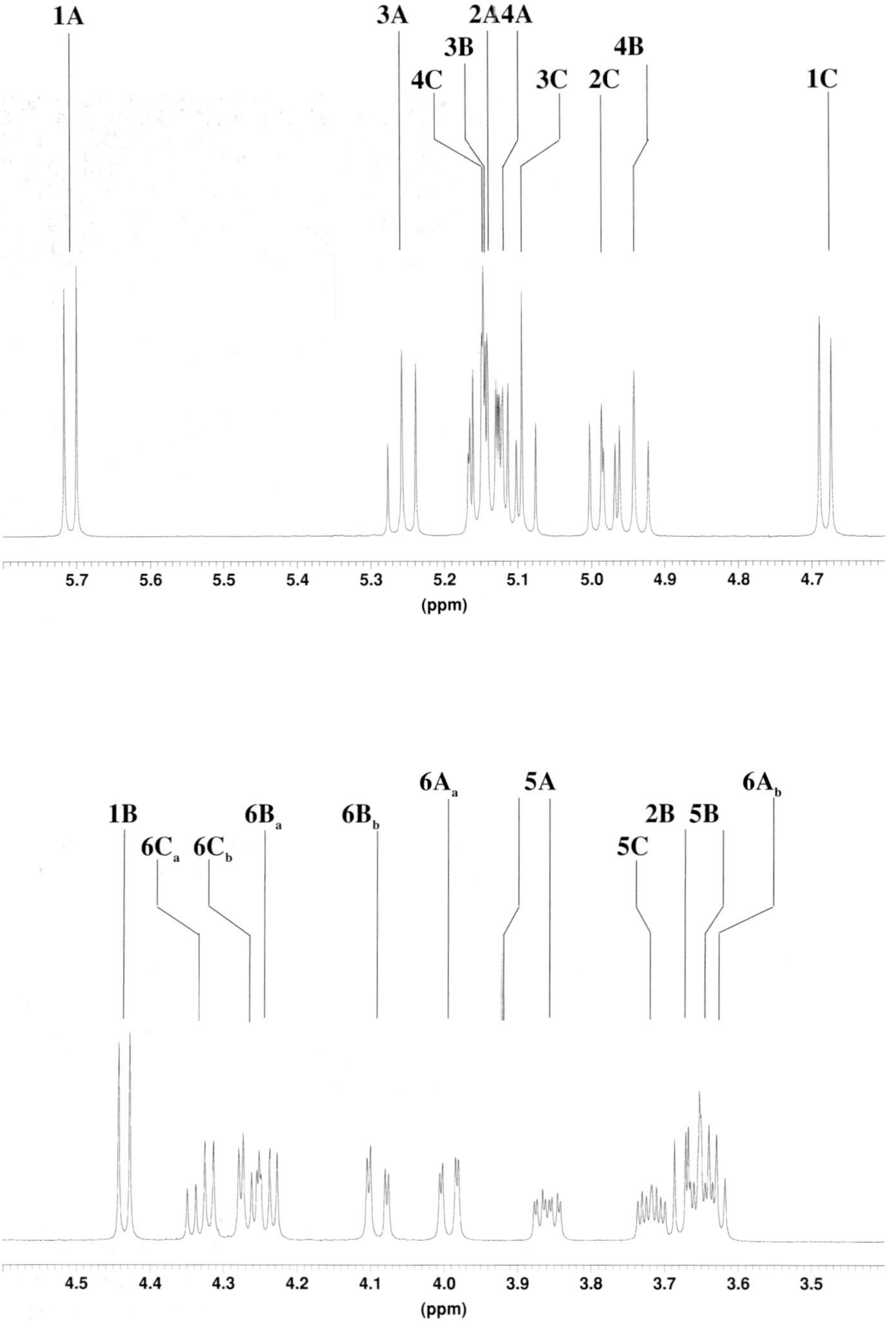

Fig. 6.2: ^{1}H-signal assignments for the peracetylated oligosaccharide.

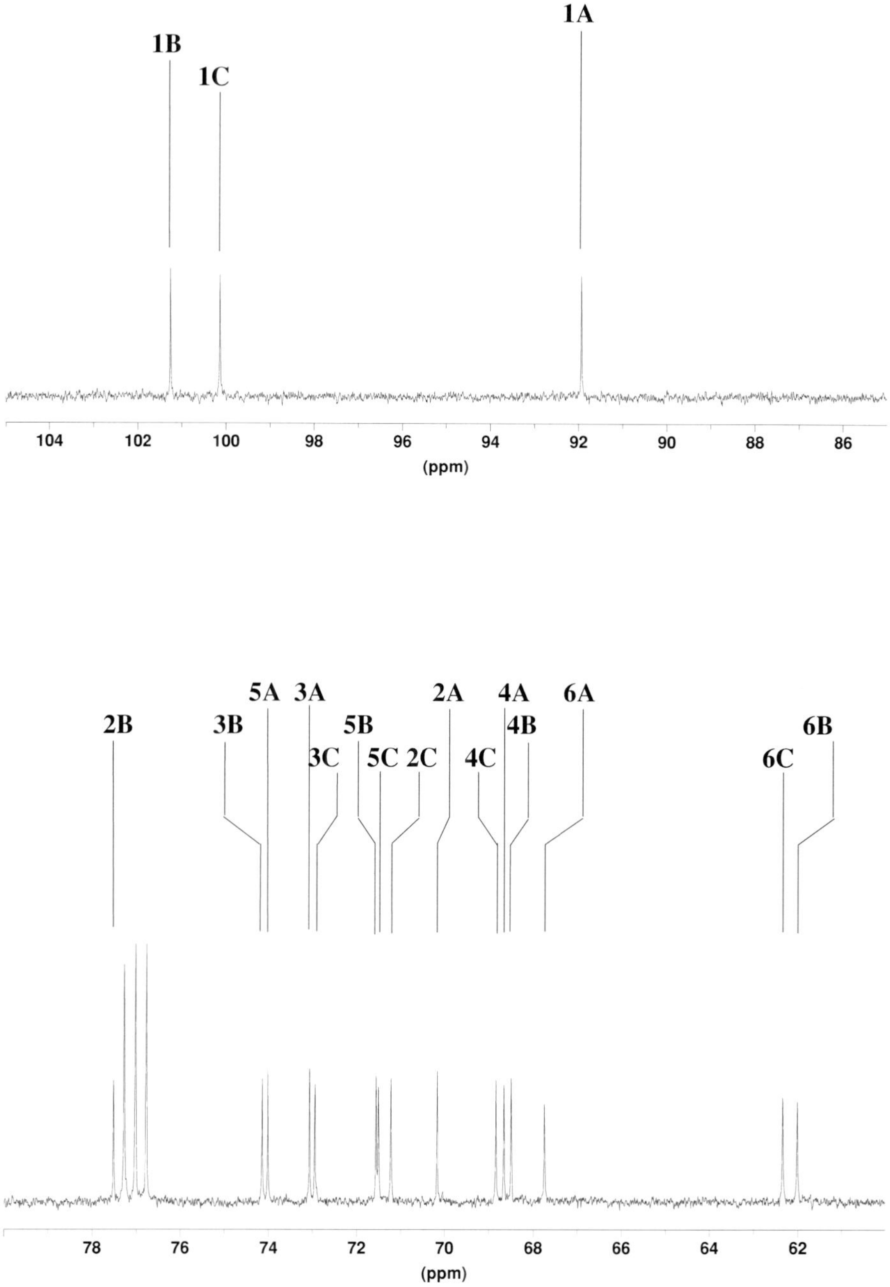

Fig. 6.3: ^{13}C-signal assignments for the peracetylated oligosaccharide.

All the above findings and conclusions are in agreement with the following structure (Fig. 6.4):

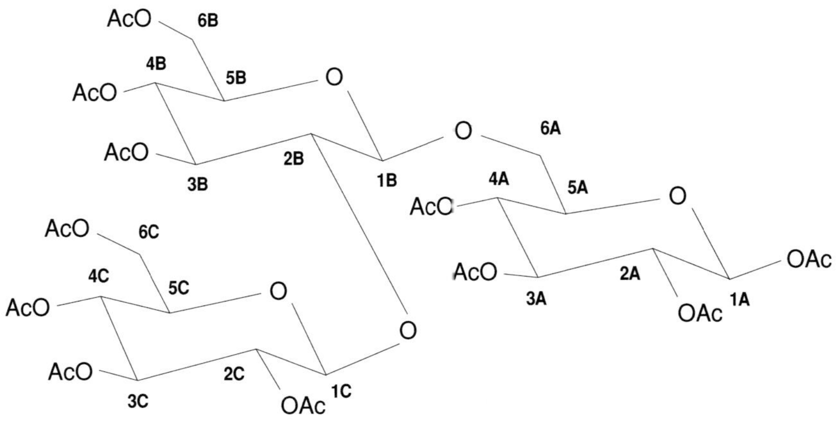

Fig. 6.4: The structure of the "unknown" peracetylated oligosaccharide.

6.5 Recommended Reading

Bremser, W., Franke, B., Wagner, H., *Chemical Shift Ranges in Carbon-13 NMR Spectroscopy*, VCH, Weinheim, 1982

Breitmaier, E., Voelter, W., *^{13}C Spectroscopy*, 3rd edition, VCH, Weinheim, 1987
^{13}C chemical shifts of carbohydrates, chapter 5.4

Pretsch, E., Clerc, J. T., *Spectra Interpretation of Organic Compounds*, VCH, Weinheim, 1997

Pretsch, E., Clerc, J. T., Seible, J., Simon, W., *Table of Spectral Data for StructureDetermination of Organic Compounds*, Springer Verlag, Berlin, 2nd edition, 1989

Recommendations for the Presentation of NMR Data for Publication in Chemical Journals, Pure Appl. Chem., 1972, *29*, 627

Presentation of NMR Data for Publication in Chemical Journals – B. Conventions relating to Spectra from Nuclei other than Protons, Pure Appl. Chem., 1976, *45*, 217

Glossary

ADC	Analog-to-Digital Converter
APT	Attached Proton Test
AT	Acquisition Time
BB	BroadBand, as in decoupling
BC	Baseline Correction
BIRD	BIlinear Rotation Decoupling
CH-COSY	Carbon-Hydrogen COrrelation SpectroscopY
COLOC	COrrelation through LOng range Couplings
COSY	COrrelated SpectroscopY
CW	Continuous Wave
DEPT	Distorsionless Enhancement by Polarization Transfer
DQ	Double Quantum
DQF	Double Quantum Filter
E/A	Echo/Antiecho selection
EM	Exponential Multiplication
FID	Free Induction Decay
FT	Fourier Transform(ation)
FTP	File Transfer Protocol
GB	Parameter for Gaussian Multiplication
GM	Gaussian Multiplication
GL	Parameter for Gaussian Multiplication
HMBC	Heteronuclear Multiple-Bond Correlation
HMQC	Heteronuclear Multiple Quantum Coherence
HSQC	Heteronuclear Single Quantum Coherence
HR	High Resolution
IFT	Inverse Fourier Transform(ation)
INEPT	Insensitve Nuclei Enhanced by Polarization Transfer
IR	Inversion Recovery
JRES	J-RESolved
LB	Line Broadening factor for exponential weighting
LP	Linear Prediction
MAPI	Mail APplication Interface
MDI	Multi Document Interface
MEM	Maximum Entropy Method
NFS	Network File Server

NMR	Nuclear Magnetic Resonance
NOE	Nuclear Overhauser Effect
NOESY	Nuclear Overhauser Effect SpectroscopY
PPM	Parts Per Million
PW	Pulse Width
QSINE	Squared SINE bell weighting function
RF	Radio Frequency
ROESY	Rotating Frame Overhauser Effect SpectroscopY
SI	SIze (total number of data points)
SINE	SINE bell weighting function
S/N	Signal-to-Noise ratio
SPC	SPeCtrum
SSB	Shifted Sine-Bell weighting factor
TD	Time domain Data points
TOCSY	TOtal Correlation SpectroscopY
T-ROESY	Transverse- Rotating Frame Overhauser Effect SpectroscopY
TPPI	Time Proportional Phase Incrementation
Z-FILTER	Pulse sandwich for elimination of signal components with dispersive phase
T_1	Longitudinal (spin-lattice) relaxation time for z-magnetization
T_2	Transverse (spin-spin) relaxation time for x,y-magnetization
tm	mixing time
tn	time domain in the n-th dimension
fn	frequency domain in the n-th dimension

Index

DATE DUE

GAYLORD			PRINTED IN U.S.A.